T0332537

FAULT COVERING PROBLEMS IN RECONFIGURABLE VLSI SYSTEMS

THE KLUWER INTERNATIONAL SERIES
IN ENGINEERING AND COMPUTER SCIENCE

VLSI, COMPUTER ARCHITECTURE AND
DIGITAL SIGNAL PROCESSING
Consulting Editor
Jonathan Allen

Latest Titles

Synchronization Design for Digital Systems, T. H. Meng
ISBN: 0-7923-9128-4
Hardware Annealing in Analog VLSI Neurocomputing, B. W. Lee, B. J. Sheu
ISBN: 0-7923-9132-2
Neural Networks and Speech Processing, D. P. Morgan, C.L. Scofield
ISBN: 0-7923-9144-6
Silicon-on-Insulator Technology: Materials to VLSI, J.P. Colinge
ISBN: 0-7923-9150-0
Microwave Semiconductor Devices, S. Yngvesson
ISBN: 0-7923-9156-X
A Survey of High-Level Synthesis Systems, R. A. Walker, R. Camposano
ISBN: 0-7923-9158-6
Symbolic Analysis for Automated Design of Analog Integrated Circuits,
G. Gielen, W. Sansen,
ISBN: 0-7923-9161-6
High-Level VLSI Synthesis, R. Camposano, W. Wolf,
ISBN: 0-7923-9159-4
*Integrating Functional and Temporal Domains in Logic Design: The False Path
Problem and its Implications,* P. C. McGeer, R. K. Brayton,
ISBN: 0-7923-9163-2
Neural Models and Algorithms for Digital Testing, S. T. Chakradhar,
V. D. Agrawal, M. L. Bushnell,
ISBN: 0-7923-9165-9
Monte Carlo Device Simulation: Full Band and Beyond, Karl Hess, editor
ISBN: 0-7923-9172-1
The Design of Communicating Systems: A System Engineering Approach,
C. J. Koomen
ISBN: 0-7923-9203-5
Parallel Algorithms and Architectures for DSP Applications,
M. A. Bayoumi, editor
ISBN: 0-7923-9209-4
Digital Speech Processing: Speech Coding, Synthesis and Recognition
A. Nejat Ince, editor
ISBN: 0-7923-9220-5
Sequential Logic Synthesis, P. Ashar, S. Devadas, A. R. Newton
ISBN: 0-7923-9187-X
Sequential Logic Testing and Verification, A. Ghosh, S. Devadas, A. R. Newton
ISBN: 0-7923-9188-8
Introduction to the Design of Transconductor-Capacitor Filters,
J. E. Kardontchik
ISBN: 0-7923-9195-0
The Synthesis Approach to Digital System Design, P. Michel, U. Lauther, P. Duzy
ISBN: 0-7923-9199-3

FAULT COVERING PROBLEMS IN RECONFIGURABLE VLSI SYSTEMS

by

Ran Libeskind-Hadas
University of Illinois, Urbana-Champaign

Nany Hasan
IBM Corporation

Jason Cong
University of California, Los Angeles

Philip K. McKinley
Michigan State University

C. L. Liu
University of Illinois, Urbana-Champaign

KLUWER ACADEMIC PUBLISHERS
Boston/Dordrecht/London

Distributors for North America:
Kluwer Academic Publishers
101 Philip Drive
Assinippi Park
Norwell, Massachusetts 02061 USA

Distributors for all other countries:
Kluwer Academic Publishers Group
Distribution Centre
Post Office Box 322
3300 AH Dordrecht, THE NETHERLANDS

Library of Congress Cataloging-in-Publication Data

Fault covering problems in reconfigurable VLSI systems / by Ran
    Libeskind-Hadas ... [et al.].
       p. cm. -- (The Kluwer international series in engineering and
computer sciences.   VLSI, computer architecture, and digital signal
processing)
       Includes bibliographical references and index.
       ISBN 0-7923-9231-0 (alk. paper)
       1. Integrated circuits--Very large scale integration--Design and
construction--Data processing.   2. Integrated circuits--Wafer-scale
integration--Design and construction--Data processing.   3. Fault
-tolerant computing.   I. Libeskind-Hadas, Ran.   II. Series.
TK7874.F38   1992
621.39'5--dc20   92-4369
                                               CIP

Printed on acid-free paper.

Printed in the United States of America

This book is dedicated to the Department of Computer Science at the University of Illinois at Urbana-Champaign.

CONTENTS

Preface   xiii

1 An Overview   1
   1.1 Introduction   1
   1.2 The Embedding Approach   3
   1.3 The Covering Approach   7
         1.3.1 Previous Work   7
         1.3.2 Physical Implementation Issues
                 in Reconfigurable Design   9
   1.4 Overview of Remaining Chapters   15

2 Fault Covers in Rectangular Arrays   19
   2.1 Introduction   19
   2.2 Admissible Assignments   21
   2.3 The Feasible Minimum Cover Problem   24
         2.3.1 Critical Sets   25
         2.3.2 An Exhaustive Search Algorithm for the
                 Feasible Minimum Cover Problem   33
         2.3.3 Experimental Results   38
   2.4 The Feasible Cover Problem   39
         2.4.1 Excess-k Critical Sets   39
         2.4.2 Experimental Results   45
   2.5 Two Reconfiguration Problems   47
         2.5.1 Reconfiguration with Shared Spares   49
         2.5.2 Reconfiguration of Programmable Logic
                 Arrays   50
   2.6 Summary   54

3 Fault Covers in Heterogeneous and General Arrays   **55**
  3.1 Introduction   55
  3.2 Fault Covers in Heterogeneous Arrays   56
      3.2.1 The Feasible Cover Problem   58
      3.2.2 The Feasible Minimum Cover Problem   62
      3.2.3 The Minimum Feasible Cover Problem   64
      3.2.4 The Feasible Cover Problem with Multiple
             Spare Arrays   71
      3.2.5 Applications of the Heterogeneous Array
             Model   74
  3.3 Fault Covers in General Arrays   76
      3.3.1 The Feasible Cover Problem   78
      3.3.2 The Feasible Minimum Cover Problem   83
      3.3.3 The Minimum Feasible Cover Problem   86
  3.4 Summary   89

4 General Formulation of Fault Covering Problems   **91**
  4.1 Introduction   91
  4.2 A General Formulation   93
  4.3 Illustrative Examples   96
  4.4 Integer Linear Programming Approach   100
      4.4.1 The General Transformation   101
      4.4.2 Experimental Results   105
  4.5 Complexity Analysis of Subcases   107
      4.5.1 The Definition of Subcases and Their
             Complexities   108
      4.5.2 Polynomial Time Algorithms   110
      4.5.3 NP-Completeness Results   112
  4.6 Summary   116

Bibliography   **119**

Index   **129**

List of Figures

1.1   A memory chip with spare rows and columns.   8

1.2   A $2^m \times 2^n$ memory chip with 2^k data bus. 10

1.3   (a) An m-bit decoder. (b) A 4-bit decoder for the address 0110. 10

1.4   A memory chip with spare rows and columns. 11

1.5   (a) A generic spare row decoder. (b) A 4-bit spare row decoder reconfigured for the address 0110. 12

1.6   Two different designs allowing the disconnection of a row containing faulty elements. 13

2.1   A reconfigurable array. 23

2.2   (a) Standard expansion of partial solution. (b) Enlarging the partial solution with an admissible assignment prior to expansion. 24

2.3   A faulty array and the corresponding critical set. 26

2.4   An array and its bipartite graph representation. 26

2.5   Illustration of Lemma 1. 29

2.6   Vertex sets used in computation of critical sets. 31

2.7   Illustration of Theorem 4. 33

2.8   Example of the Min-Cover algorithm. 36

2.9   (a) The graph G and related sets. (b) The graph $G(v)$ and matching $\mathcal{M}(v)$. 42

2.10 Two arrays sharing spare columns. 49

2.11 A redundant Programmable Logic Array. 51

3.1   A reconfigurable array with multiple types of elements. . . 56

3.2   Three faulty heterogeneous arrays. 58

3.3   Multigraph for feasibility problem. 60

3.4   Construction of G_3 from G_1 and G_2.   68
3.5   Construction of arrays from a graph.   70
3.6   Multiple spare array problem.   72
3.7   Reconfiguration with shared spares.   75
3.8   Reconfigurable stack of arrays.   75
3.9   Example of the general array model.   77
3.10 Construction of array A from arrays $A_1, \ldots, A_t$.   87

4.1   The chip of Example 4.1.   92
4.2   The generalized bipartite graph for Example 4.1.   96
4.3   (a) A faulty homogeneous array. (b) The corresponding
      graph.   98
4.4   (a) Nine arrays with shared spares. (b) The corresponding
      generalized bipartite graph.   99
4.5   (a) An array with interstitial redundancy. (b) The corre-
      sponding generalized bipartite graph.   100
4.6   (a) Interstitial array with two types of spares. (b) The
      corresponding generalized bipartite graph.   101
4.7   Splitting vertices u, v, and w.   110
4.8   Network construction for problem 13.   113
4.9   An example of the construction in Theorem 16.   117

List of Tables

2.1   Experimental results for the feasible minimum cover problem. 40

2.2   Experimental results. 48

4.1   Experimental results for homogeneous arrays. 106

4.2   Experimental results for homogeneous arrays. 106

4.3   Experimental results for arrays with shared spares. 107

4.4   Experimental results for interstitial arrays with two types of spares. 108

4.5   Sixteen subcases of the general formulation. 109

4.6   Experimental results for problem 13. 113

Preface

This monograph presents our recent research on reconfiguration problems for fault-tolerance in VLSI and WSI systems. Reconfiguration problems can generally be classified as either embedding problems or covering problems. We have directed our efforts at covering problems since this approach has been used most widely in commercial applications. Our goal is to provide efficient algorithms for tractable covering problems and to provide general techniques for dealing with an even larger number of intractable covering problems. We hope that the results and techniques described here will be useful as guides for other researchers working in this area. We begin by investigating algorithms for the reconfiguration of large redundant memories. We then describe a number of more general covering problems and analyze the complexity of these problems. Finally, we consider a general and uniform approach to solving a wide class of covering problems.

Work reported in this monograph was partially supported by the National Science Foundation under grants MIP 8703273 and MIP 8906932.

Chapter 1

An Overview

1.1 Introduction

The area of fault-tolerant computing has become increasingly important with the growth in complexity of computer hardware and software. Hardware issues range from the design and analysis of fault-tolerant circuits to large fault-tolerant computing systems. Software issues range from the design and analysis of fault-tolerant data structures to large fault-tolerant software systems. While many problems are being actively studied in both the hardware and software domains, efforts at designing fault-tolerant VLSI and WSI systems have been of particular interest in the last decade. This is due to both the very large number of components in such designs, and thus the high likelihood of faults, as well as to the regular structure of many of these architectures, which makes them particularly well-suited for fault-tolerant design.

Fault-tolerant designs have been proposed for a wide variety of VLSI architectures, such as arrays of processors, binary trees of processors, multipipelines, and random access memories. In general, such fault-tolerant systems contain more elements than are actually required for the system to operate. In addition, reconfiguration hardware is provided to allow some elements to be replaced by others. Such systems are referred to as *reconfigurable systems*. For a given system with some faulty elements, the *reconfiguration problem* is that of determining how the faulty elements can be replaced to enable the system to function as required.

There are two distinct objectives in the design of fault-tolerant VLSI systems. One objective, known as *yield enhancement*, is to design systems that can be reconfigured by the manufacturer, if faulty elements are detected in the testing phase. The reconfiguration hardware used to replace faulty elements by surplus elements in this case are generally *hard switches*, which are electrically-programmed antifuses or laser-programmed switches that cannot be subsequently reprogrammed. Because of the nature of the reconfiguration circuitry, this type of reconfiguration is known as *static reconfiguration*. Static reconfiguration for yield enhancement has been employed widely, especially in the design of random access memories.

Another objective is to design systems that can be reconfigured when faults occur during the operation of the system so that the system *reliability* is improved. This type of reconfiguration, known as *dynamic reconfiguration*, usually requires *soft switches* that may be reprogrammed during the lifetime of the system in order to accommodate various fault patterns. These switches are generally controlled by writable memory cells, such as SRAM or EEPROM cells. Reconfiguration may be carried out by the circuit itself or by a host computer. The dynamic reconfiguration approach is particularly useful in arrays of processors and other devices in which the elements have a significant probability of failure during operation.

There are two distinct approaches to reconfiguration that are applicable in both the static and dynamic cases. In the *covering* approach, a number of spare elements are available and each spare element may be used to replace some set of regular elements that may become faulty. In this approach, the reconfiguration problem becomes that of finding an assignment of spare elements to faulty regular elements such that the spare elements replace all the faulty elements. Such an assignment is called a *covering assignment*. In the *embedding* approach, no distinction is made between regular and spare elements. The system contains more elements than are required so that if some of the elements fail, the desired target architecture can be mapped into, or *embedded* in, the remaining non-faulty elements.

While the embedding approach has been proposed for a number of architectures, the amount of reconfiguration circuitry required is considerably more than that of the covering approach. Consequently, the covering approach has been more widely used in commercial products. Several manufacturers have adopted the covering approach in the reconfiguration of random access memory chips. Success with the covering approach suggests that it may soon become a viable reconfiguration strategy for other types of architectures. Therefore, after briefly reviewing research related to the embedding approach, we shall restrict our attention to a number of covering problems in reconfigurable VLSI and WSI systems. We present algorithms for several covering problems, study the computational complexity of a number of very general covering problems, and propose a new way to formulate most covering problems in a way that they can be solved optimally by the method of integer linear programming.

In Section 1.2 we present an overview of the embedding approach. In Section 1.3 we describe the covering approach in detail, and in particular how the covering approach is used in the reconfiguration of memory chips. We also examine some of the physical implementation issues associated with these reconfigurable memories. In Section 1.4 we present an overview of the results and contributions of this book.

1.2 The Embedding Approach

Although the objective of this book is to study the covering approach, in this section we give a brief overview of research related to the embedding approach. The discussion here is intended only to familiarize the reader with some of the results in this area. The remainder of this book does not assume any of the results from this section.

In the embedding approach, which is also referred to as *processor switching* [7], no distinction is made between regular elements and spare elements. The objective is to *embed* the desired target topology in the remaining non-faulty elements. Embedding has also been called *index-mapping* because the problem can be viewed as mapping a particular topology onto another topology consisting of the non-faulty components on the chip or wafer.

Linear Arrays. The simplest target topology is the linear array. Koren [42] described how linear arrays can be embedded in two-dimensional arrays with faulty processors. Tyszer [88] gave an algorithm for embedding a linear array in a chordal ring. Unlike most other approaches, Tyszer's approach accommodates both element faults and link faults and can tolerate up to $m - 1$ element and link failures, where m is the degree of each element in the chordal ring. Shombert and Siewiorek [74] described a system of *roving spares* for linear arrays, primarily for the purpose of temporarily taking elements out of service in order to run diagnostic programs. Finally, Gupta *et al.* [23] gave an algorithm and implementation for extracting multiple pipelines from two-dimensional arrays with faulty elements.

Two-Dimensional Arrays. The subject of fault tolerance in two-dimensional arrays has been studied extensively. The CHiP architecture [80] was one of the first designs using embedding in a two-dimensional array. The idea of embedding regular structures was discussed in [80], although no specific reconfiguration algorithms were given there. In another approach, called *direct reconfiguration* [70], surplus elements are located on two sides of the array (e.g., right and bottom). A faulty element is either replaced by the element to its right or left, and replacement propagates until a surplus element is used.

An alternative approach, called *fault-stealing*, was proposed by Sami and Stefanelli [69]. In *fixed-stealing*, rows are considered from top to bottom. The rightmost faulty element in a row is shifted right; all other faulty elements are shifted down. A more flexible scheme, called *variable-stealing*, allows any of the faulty elements in a row to be shifted to the right. In this case, choosing a faulty element whose lower neighbor is also faulty decreases the probability of reaching a fatal failure [70]. Finally, allowing element (i, j) to steal from element $(i + 1, j + 1)$ is called *complex-stealing* [68], and increases the probability of chip survival at the cost of more complex circuit design.

Kung *et al.* [45] extended the fault stealing approach for two-dimensional arrays in two ways. First, to save space, their model allows only a single track for communication links between rows and columns. Second, in addition to a centralized reconfiguration algorithm, a distributed version of the algorithm is also given. The distributed algorithm is intended to run in real time to enhance reliability during operation. Roychowdhury *et al.* [67] have given a polynomial-time reconfiguration algorithm for this model.

Another approach to fault stealing is the *full use of suitable spares* (FUSS) strategy [8], in which extra columns of elements augment the array. This method is based on a *surplus vector*, which contains an entry for each row of the array. The entry for a row indicates a surplus or deficiency of extra elements in that row. Using the surplus vector, elements are reassigned so as to leave each logical row with the desired number of elements.

Some of the research on fault tolerance in two-dimensional arrays is specifically oriented toward systolic arrays. Reconfiguration of systolic arrays is usually based on the idea of *data flow paths*. A data flow path is a linear set of elements through which data may pass from the first element to the last element. A *cut* is a set of elements that, if bypassed, leads to an array with one fewer data flow paths. In Koren's model [42], entire rows and columns of the array must be bypassed as a unit, that is, cuts constitute only rows and columns. In Kung and Lam's model [44], however, elements in cuts need not be from the same row or column. These two approaches were compared in [48] where it was shown that the redundancy required in the former approach is not polynomially bounded, whereas in the latter approach it is. Li *et al.* [49] gave an algorithm for restructuring systolic arrays in the presence of faults. The algorithm is based on the identification of the paths required to implement the systolic computation.

Binary Trees. The binary tree topology is particularly well suited for computations in a number of applications. Two basic problems related to fault-tolerance have been studied for VLSI implementations of binary trees. The first problem is to design a topology that will remain connected in the presence of element or link failures. An example is the X-tree topology [13]. In related work, Bertossi and Bonuccelli [6] introduced the Cousin-Connected Tree (CCT), in which each element is connected to one of its two cousins. It was shown that, in the presence of processor or link failures, this topology can be reconfigured into a ternary tree, a topology suitable for solving linear programming problems.

The second problem seeks to embed a complete binary tree in a given topology. The problem of reconfiguring the non-faulty elements into a binary tree was shown to be NP-complete [VaFu82]. A design that tolerates single faults in symmetric non-homogeneous trees was proposed by Hayes [30]. Nodes at different levels are assumed to provide different types of computing capabilities. Kwan and Toida [47] extended Hayes' approach to include asymmetric trees that can tolerate one fault and binary symmetric trees that can tolerate two faults.

In the SOFT approach [55], additional elements are located at the leaves of the tree and are shared by subtrees with spare links connecting siblings in the tree. This approach is also capable of tolerating failures of links and switches.

Other approaches include the Cluster-Proof scheme [35, 36], which seeks to tolerate clustered faults, and a scheme proposed by Dutt and Hayes [14] that can tolerate any combination of k faults.

Other Target Topologies. The embedding approach has been studied for several other target topologies. For example, Banerjee [3] proposed a fault-tolerant version of the cube-connected cycles (CCC) architecture [62], which can tolerate failures of any number of processors and links in a single cycle. Lin and Wu [51] presented a fault-tolerant mapping scheme for a configurable multiprocessor system using multistage switching networks.

The Diogenes Approach. Embedding schemes can usually be grouped according to the target topology. However, the Diogenes approach [66] is applicable in virtually any point-to-point target topology. In the Diogenes approach, processor elements are fabricated along a bundle of unconnected wires and controlling switches. Fault-free elements are then connected to form the desired topology, bypassing faulty elements. This approach can been formalized in terms of hypergraphs [65]. The two most important virtues of this scheme are its simplicity and its applicability to a wide assortment of target topologies, including linear arrays, two-dimensional arrays, trees, and n-cubes. A disadvantage of the Diogenes approach is the potential for long interconnection lines, even in the absence of faults, which can increase both signal propagation delay and the susceptibility of the chip to interconnection faults [75]. This situation necessitates a relatively slow system clock rate. Varman and Ramakrishnan [90] addressed this problem by placing buffers in long communication paths such that all data will travel the same physical distance, permitting higher clock rate and thus higher system throughput.

1.3 The Covering Approach

While the embedding approach has many attractive features, the amount of reconfiguration circuitry required is often prohibitively expensive in terms of chip area and performance. This is particularly true for systems comprising a large number of relatively simple elements. The covering approach has been proposed to reduce chip area and improve performance in such architectures. Examples of such architectures are memories, certain types of processor arrays, and binary trees of processors.

In the covering approach, the elements in the system are divided into two classes: Regular elements and spare elements. Spare elements may be used to replace faulty regular elements. However, regular elements cannot be used to replace other regular elements. The amount of reconfiguration circuitry is significantly reduced in this approach since switches are only needed between regular elements and spare elements and because the number of spare elements needed is generally quite small.

We begin this section by describing some of the previous work on applications of the covering approach for such architectures. Next, we describe in some detail the reconfiguration hardware used in fault-tolerant memories, the architecture that has benefited most from the covering approach.

1.3.1 Previous Work

By far the most significant example of the covering approach is for rectangular arrays of memory elements or processor elements in which spares are configured as entire rows and columns. Each faulty element in the array is replaced by replacing the entire row in which the faulty element resides by a spare row or by replacing the entire column in which the faulty element resides by a spare column. An array containing faulty elements will subsequently be referred to as a *faulty array*. We let SR denote the number of spare rows and SC denote the number of spare columns. An example of a faulty array is shown in Figure 1.1 in which there are four faulty elements represented by darkened squares, $SR = 2$,

and $SC = 1$. A covering assignment for the array in Figure 1.1 assigns
the two spare rows to rows 1 and 4 and the spare column to column 2,
or assigns the two spare rows to rows 2 and 3 and the spare column to
column 4.

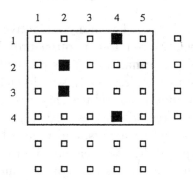

Figure 1.1: A memory chip with spare rows and columns.

Early proposals for this type of covering were made by Tammaru
and Angell [84] in 1967 and by Chen [9] in 1969 for large-scale integrated semiconductor memories. Since then, a number of researchers
have studied the amount of yield-enhancement achievable by this method
[9, 20, 81, 82, 83, 84], the type of switching hardware needed to implement
this type of reconfiguration [52, 56, 78, 79], and algorithms and heuristics
for finding a covering assignment [11, 26, 32, 46, 54, 86]. Many industrial
laboratories have reported success with this method, and many memory
manufacturers are using this approach in their products. The covering
approach has also been proposed for systolic arrays [40, 57], for NASA's
MPP Computer [4], as well as in a number of other rectangular arrays of
processors. Although the strategy of replacing entire rows and columns
suffers from low spare utilization, this technique minimizes the complexity of the reconfiguration circuitry since switches are provided only for
spare rows and columns instead of for each individual element.

The covering approach has also been used in the reconfiguration of
binary trees of processors. In contrast to the embedding approach described in the previous section, in the covering approach the tree is partitioned into local regions, each with its own set of spare processors. A
faulty processor can be replaced by any of the local spares linked to it.

Raghavendra *et al.* [63] presented an approach in which one spare element, along with spare links, is available for each level of the tree in order to protect against failures at that level. A generalization in which one spare is provided per 2^i elements, for any i, was also given, but requires a larger number of spare links [75]. Another local reconfiguration scheme for binary trees was given by Hassan and Agrawal [28]. One spare element is provided for every three-element subtree, or module, implying that only one fault per module is tolerable. Extensions with lower redundancy ratios were also discussed. Advantages include simple, localized redundancy, area-efficient VLSI layout, and applicability to yield enhancement in addition to reliability [75]. The major drawback of this approach is that trees with clustered faults tend not to be repairable [36].

Another example of the covering approach is the *interstitial redundancy* scheme proposed by Singh [76]. In this approach, spare elements are placed in interstitial sites between regular elements, and spares are connected to their nearest regular neighbors. This method is well-suited for large arrays of complex processor elements; the extra cost of the reconfiguration circuitry is balanced by the high probability of faulty processors and the high utilization of the spare processors. Interstitial redundancy has been proposed for a number of designs, including the 3-D computer developed by Hughes Research Laboratories [94]. This approach would not be suitable for memories, however, since the reconfiguration circuitry would consume a large fraction of the total circuit area. The complexity of reconfiguration algorithms for this approach has been studied in [24].

1.3.2 Physical Implementation Issues in Reconfigurable Design

In this subsection we describe the reconfiguration circuitry and other hardware issues related to the design of reconfigurable random-access memories.

Circuit Design for Reconfigurable Memories

A memory chip is essentially a two dimensional array of memory cells. Each row of the memory array is called a *word line* and each column is called a *bit line*. For a $2^m \times 2^n$ memory chip with a 2^k-bit data bus, the row address bus is m-bits wide, and there are 2^m m-bit row decoders. The column address bus is $(n-k)$-bits wide, and there are 2^{n-k} $(n-k)$-bit column decoders. See Figure 1.2. An m-bit decoder can be implemented using a NOR gate with m inputs. Figure 1.3a shows an m-bit decoder using NOR gate implementation, and Figure 1.3b shows a 4-bit decoder for the address 0110.

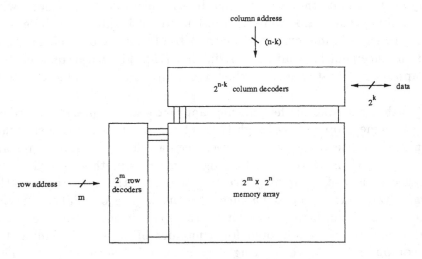

Figure 1.2: A $2^m \times 2^n$ memory chip with 2^k data bus.

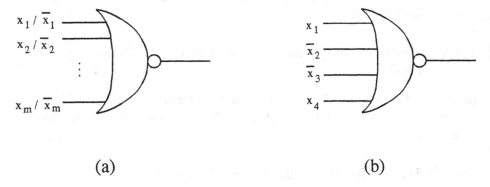

(a)   (b)

Figure 1.3: (a) An m-bit decoder. (b) A 4-bit decoder for the address 0110.

In a reconfigurable memory chip, spare rows and columns are included in the chip together with redundant row decoders and column decoders. This is illustrated in Figure 1.4. We assume that each spare row (column) can be used to replace any of the regular rows (columns) in the chip [1]. We shall describe the circuitry used for row redundancy; the circuitry used for column redundancy is similar [2].

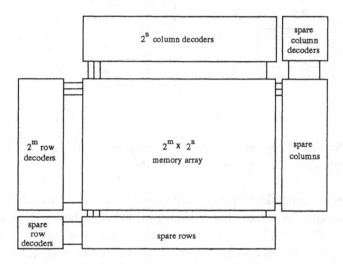

Figure 1.4: A memory chip with spare rows and columns.

The memory cells in the spare rows are identical to those in the regular array. However, the spare row decoder is slightly different from the regular row decoder. An m-bit spare row decoder can be implemented using a $2m$-input (instead of an m-input) NOR gate. There is a fusible link at each of the $2m$ inputs of the NOR gate. For the i-th bit of the row address bus, both x_i and $\bar{x}_i$ are connected as inputs of the NOR gate. Therefore, a spare row decoder has a default output of 0. Figure 1.5a shows a generic m-bit spare row decoder. If a spare row is used to replace an original row at address $a_1a_2\ldots a_m$, we reconfigure the fusible links at the inputs of the spare decoder as follows: If $a_i = 0$, we break the fusible link for $\bar{x}_i$; if $a_i = 1$, we break the fusible link for x_i. It is easy to verify that after such reconfiguration, the spare row decoder will decode the address $a_1 a_2 \ldots a_m$ correctly. For example, Figure 1.5b shows a 4-bit spare row decoder reconfigured for the address 0110.

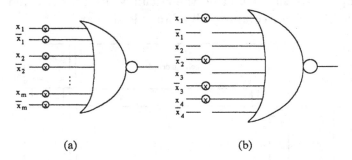

(a)   (b)

Figure 1.5: (a) A generic spare row decoder. (b) A 4-bit spare row decoder reconfigured for the address 0110.

In a reconfigurable memory chip, we also need the capability to disconnect a regular row from the memory array when faulty elements occur in the row. The simplest solution is to have a fusible link at the output of each regular row decoder. Figure 1.6a shows the schematic of such a design. If we wish to disconnect a row from the memory array, we can simply break the fusible link for the output of the corresponding row decoder so that the row will never be selected. This approach was used in the 64K DRAM memory chip manufactured by AT&T Bell Labs [79]. Although such a design is simple and will introduce very little extra delay in the memory access time [3], it uses a large number of fusible links. For a $2^m \times 2^n$ memory array with SR spare rows, we need $2^m + 2m \cdot SR$ fusible links, where 2^m fusible links are used for the regular row decoders and $2m \cdot SR$ fusible links are used for the spare row decoders. An alternative method is to generate a RRD (Regular Row Disable) signal whenever a spare row is selected. The RRD signal is available as input to each regular row decoder. Figure 1.6b shows the schematic of such a design. When a spare row decoder is selected, the RRD signal becomes high so that no regular row is selected. (Recall that each decoder uses a NOR gate in this design.) Such an approach was used in the 16K SRAM

---

[3]This extra delay is due to the extra capacitance introduced by the fusible links.

reconfigurable memory chip manufactured by Intel [41]. The advantage of this design is that it uses far fewer fusible links. For a $2^m \times 2^n$ memory array with SR spare rows, only $2m \cdot SR$ fusible links are needed for the spare row decoders. However, in this design extra delay is introduced for memory access time. This is true because when a spare row is read, the data is not available until the regular rows are completely disconnected after the RRD signal is generated to disable the regular row decoders. Usually, the increase in access time is relatively small. For example, for the 16K SRAM reconfigurable memory chip manufactured by Intel [41], the access time was 40ns before adding redundancy and it became 43ns afterwards. If minimizing access time is critical to the design, extra circuitry may be used to eliminate the delay due to the time for disconnecting the regular rows. This may be done by introducing another set of data latches at the output of the memory chip. Interested readers may refer to the design of reconfigurable memory chips manufactured by IBM [20] for further details.

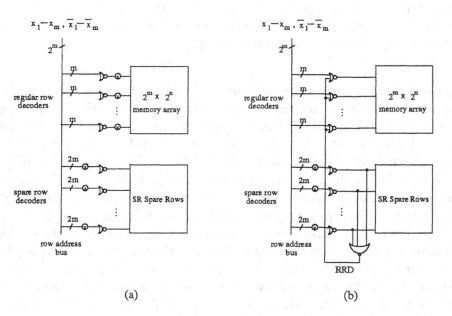

Figure 1.6: Two different designs allowing the disconnection of a row containing faulty elements.

Reconfiguration Techniques

When a regular word or bit line is replaced by a spare line, the address of the faulty line must be stored. This information can be stored by either fusible links or dynamic storage elements (such as latches). Dynamic storage elements have the advantage of being reprogrammable. However, the information in the dynamic storage elements is volatile, meaning that the information stored will be lost each time the power is turned off. Therefore, if dynamic storage elements are used to store the addresses of faulty lines, stand-by power must be supplied to the circuit or the information must be backed-up on some permanent storage media (such as disks) and re-loaded each time the chip is reset. Because of the volatility of dynamic storage elements, they are not commonly used in reconfigurable chips. Instead, most reconfigurable chips use fusible links, which are non-volatile but also not reprogrammable. In general, there are two kinds of fusible links, laser fusible links and electrically fusible links. Commercial reconfigurable memory chips are split about evenly between those using laser fusible links and those using electrically fusible links. In the remainder of this section, we shall discuss each of these two types of repair techniques in more detail.

Laser Programming Techniques. Laser fusible links are used in many reconfigurable memory chips, such as the 64K DRAM and the 256K DRAM from AT&T Bell Labs [5, 79], the 4K × 8 SRAM and the 8K × 8 SRAM from Intel [15, 77], the 256K DRAM and the 1M DRAM from Mostek [39, 87], and the 1M DRAM from Mitsubishi [43]. Usually, the laser fusible links are implemented in the polysilicon layer. A laser and laser beam positioning system are needed to carry out the reconfiguration. The repair process consists of the following steps for each target laser fusible link: Compute the absolute coordinates of the target link on the chip, position the laser beam at the target link according to its coordinates, and use the laser beam to blow the target link. These steps are usually controlled by a real-time computer. For example, in the 64K reconfigurable DRAM from AT&T Bell Labs, laser fusible links are 3 μm wide and 14 μm long, composed of heavily doped polysilicon. These links are spaced at least 9.5 μm apart from center to center. The effective laser spot diameter is about 7-8 μm and the accuracy of the beam positioning system is about ± 1 μm. A single laser pulse of approximately 50-ns duration is used to blow a target laser fusible link. A real-time minicomputer is used to compute the absolute coordinates of a target link and to position the laser beam at the target link. The repair process usually takes only a few seconds.

Electrical Programming Techniques. Electrically fusible links are also commonly used in reconfigurable memory chips, such as the 16K SRAM from Intel [41], the 256K DRAM from Fujitsu [58], the 32K $\times$ 8 DRAM and the 1M DRAM from IBM [19, 37], and the 64K $\times$ 8 EPROM from AMD [64]. Electrically fusible links can be further divided into electrically metal fusible links such as the ones used in the 32K $\times$ 8 DRAM and the 1M DRAM from IBM and electrically poly fusible links such as the ones used in the 16K SRAM from Intel [41] and the 64K $\times$ 8 EPROM from AMD [64]. These electrically fusible links are blown by passing high currents through the circuit. The advantage of using electrically fusible links is that no expensive hardware such as the laser beam positioning system is required. Reconfiguration is also easier since the address of a faulty row can simply be applied to the corresponding address pads in order to activate a spare row. However, extra circuitry must be included to route the high current to the appropriate electrically fusible links. Nevertheless, the resulting increase in chip area is usually acceptable. For example, the die area of the 16K SRAM from Intel was $40,000mil^2$ before adding redundancy. It became $42,582mil^2$ after incorporating redundancy (the increase also included the overhead of three spare rows). The high-current for blowing the target links is provided through a special pad on the chip, which is connected to a high-voltage supply. Usually, this pad is only used during the repair process and it is later grounded to prevent inadvertent reconfiguration.

1.4 Overview of Remaining Chapters

In the next three chapters we shall consider three issues related to the covering approach to reconfiguration problems. We begin by examining in detail the problem of reconfiguring a rectangular array with a fixed number of spare rows and spare columns. We have already noted that this reconfiguration problem arises in the context of reconfigurable memories as well as a number of reconfigurable processor arrays. The problem of reconfiguring an array with a given fault pattern and a given number of spare rows and columns was shown to be NP-complete by Kuo and Fuchs [46]. Consequently, a number of researchers have designed heuristics and exhaustive search algorithms for this problem. While heuristics tend to be very fast, a large number of problem instances can be found for which heuristics are unable to find solutions, even when they are known to exist. On the other hand, exhaustive search algorithms always find a solution if one exists, but in the worst case require exponential running time. In Chapter 2 we present some new techniques which enable us to significantly reduce the running time of exhaustive search algorithms for this problem. The techniques are based on properties of bipartite graphs and can be used to improve the efficiency of virtually all existing exhaustive search algorithms for this problem.

In Chapter 3 we describe generalizations of the reconfiguration problem studied in Chapter 2. Specifically, we consider arrays containing elements of different types. Since the elements are of different types, the rows and columns, and thus the spare rows and spare columns, are also of different types. We consider the problem of finding a covering assignment in a faulty array, showing that this reconfiguration problem can be solved in polynomial time under certain constraints but becomes NP-complete in the general case. These reconfiguration problems are shown to be relevant to a number of applications such as in the design of 3-dimensional memory arrays and memories with shared spares. The model is then further generalized and related complexity problems are discussed.

In Chapter 4 we present a general formulation of fault covering problems using generalized bipartite graphs. This formulation captures a large class of the possible relationships between faulty elements and spare elements in reconfigurable chips. We show how a reconfiguration problem in the general formulation can be transformed into an integer linear programming problem and can then be solved optimally. Although the integer linear programming problem is NP-complete, it is a well-known optimization problem that has been studied extensively. Experimental results indicate that even relatively large reconfiguration problems formulated in this way can be solved in a reasonable amount of computation time. Finally, we use this general formulation to help identify the computational complexity of several classes of covering problems.

Chapter 2

Fault Covers in Rectangular Arrays

2.1 Introduction

The covering approach is used most frequently in rectangular arrays with spare rows and columns. Several examples of such architectures, including reconfigurable random access memories (RRAMs) and processor arrays, were described in Chapter 1. Recall that such architectures comprise an $m \times n$ array with SR spare rows and SC spare columns in which each faulty element is replaced by replacing the entire row in which it resides by a spare row or by replacing the entire column in which it resides by a spare column. As we have mentioned in Chapter 1, the problem of finding a covering assignment for a faulty array is NP-complete [46]. Consequently, efforts in this area have been in the design of heuristics and exhaustive search algorithms. Heuristics can generally find solutions in a short amount of time, but may not always find solutions when they exist. In contrast, exhaustive search algorithms can always find solutions when they exist, but may require exponential running time in the worst case. In this chapter we propose new techniques that can significantly reduce the running time of exhaustive search algorithms for many of these problems.

To facilitate our discussion, we begin by introducing some notation and definitions. A *solution* to an instance of a covering problem is a covering assignment for that problem instance, that is, an assignment of spare elements to faulty elements that repair all the faulty elements. A *partial solution* is an assignment of spare elements to faulty elements that repair some, but not necessarily all faulty elements. In the case of a rectangular array, a *line* is a row or column of the array and a *cover* is a set of lines that contain all the faulty elements in the array. A set of lines is said to be *feasible* if it includes at most SR rows and SC columns. Thus, a feasible cover is a set of at most SR rows and SC columns that contain all the faulty elements in the array. Since all spare rows are identical and all spare columns are identical, a covering assignment for a rectangular array corresponds to a feasible cover for the array. The problem of finding a feasible cover for a given array is known as the *feasible cover problem.*

In some cases, the cost of reconfiguring an array is proportional to the number of spare rows and columns used. In such cases, a feasible cover containing the smallest number of lines is desirable. The *cardinality* of a cover is the number of rows and columns in the cover. A *minimum cover* is a cover with minimum cardinality. A *feasible minimum cover* is a minimum cover that is also feasible, that is, a minimum cover using at most SR rows and SC columns. The problem of finding a feasible minimum cover for a given array is known as the *feasible minimum cover problem* [26] [1].

We have noted that the feasible cover problem is NP-complete [46]. In addition, the feasible minimum cover problem was shown to be NP-complete by Shi, Chang, and Fuchs [73]. Consequently, fast heuristics and exhaustive search algorithms have been proposed for these problems. Heuristics have been studied in [11, 17, 32, 46, 54, 86, 92], but a large number of problem instances have been found for which these heuristics fail to find a solution when solutions are known to exist. Similarly, exhaustive search algorithms have been considered in [11, 26, 46, 54]. The primary advantage of exhaustive search algorithms is that they always

---

[1]Although the feasible minimum cover problem is a special case of the feasible cover problem, search algorithms for the feasible minimum cover problem tend to be simpler and faster than search algorithm for the feasible cover problem. Moreover, experimental results indicate that most arrays that have a feasible cover also have a feasible minimum cover. An experiment was conducted on 1000 1024 × 1024 randomly generated faulty arrays with clustered faults. Among the arrays that had feasible covers, the fraction of arrays that also had feasible minimum covers ranged from 63% to 97%, depending on the density of the faulty clusters.

find a solution if one exists. Unfortunately, the inefficiency of existing search algorithms has rendered them impractical in many applications.

In this chapter we introduce the concept of *admissible assignments* and show how such assignments can be used to substantially improve the running times of exhaustive search algorithms for the feasible minimum cover problem and the feasible cover problem. In Section 2.2 we describe the admissible assignment approach in detail. In Section 2.3 we show how admissible assignments can be found for the feasible minimum cover problem. We then give a search algorithm for the feasible minimum cover problem using admissible assignments and provide experimental results comparing the performance of our algorithm with a similar algorithm that does not use admissible assignments. In Section 2.4 we show how admissible assignments can be found for the feasible cover problem. We describe a search algorithm based on these admissible assignments and compare the performance of this algorithm with that of existing algorithms for this problem. In Section 2.5 we show how the admissible assignment approach can be used to improve exhaustive search algorithms for several related reconfiguration problems.

2.2 Admissible Assignments

An effective way to organize an exhaustive search is to perform the search incrementally. An *incremental exhaustive search algorithm* maintains a list of partial solutions. The algorithm selects some partial solution P from this list and *expands* P by augmenting it with additional assignments of spare elements to faulty elements, resulting in a larger partial solution P'. This process is repeated until a solution is found or the entire search space is exhausted. A solution, S, is said to be *derivable* from partial solution P if S can be found by repeated expansion of P.

Most incremental search algorithms for reconfiguration problems share a similar underlying structure: A list of partial solutions is maintained with the property that every solution to the reconfiguration problem is derivable from at least one partial solution on the list. At each iteration some partial solution, P, is selected for expansion. This partial solution is expanded into one or more new partial solutions in such a way that any solution to the reconfiguration problem derivable from P is also derivable from one of the new partial solutions. The partial solution, P, is then removed and the expanded solutions are added to the list of partial solutions. The expansion process continues until a complete solution is found or the list becomes empty, in which case no solution exists.

Recall that there is a correspondence between feasible covers and covering assignments for the feasible cover problem. Thus, partial solutions can be represented as feasible sets of lines that contain some faulty elements in the array and solutions correspond to feasible covers. The underlying structure of most search algorithms for the feasible cover problem or the feasible minimum cover problem is as follows: The algorithm maintains a list of partial solutions, initially containing only the empty partial solution consisting of no rows or columns. At each iteration of the search algorithm, some partial solution, P, is selected and removed from the list. A line, ℓ, which contains one or more faulty elements not yet replaced by the lines of P, is selected. Next, two new partial solutions, P_{include} and P_{exclude}, are created: P_{include} is formed by including the line ℓ in the partial solution P, and P_{exclude} is formed by excluding the line ℓ from the partial solution P. Observe that by excluding ℓ from the new partial solution P_{exclude}, we are forced to include all other lines containing faulty elements in line ℓ. These lines are therefore included in P_{exclude}. For example, for the array shown in Figure 2.1, assume that $P = \emptyset$. If $\ell = \{\text{row 1}\}$ is selected, then $P_{\text{include}} = \{\text{row 1}\}$ and $P_{\text{exclude}} = \{\text{column 1}, \text{column 4}, \text{column 6}\}$. If a new partial solution is not feasible then it is discarded since the partial solutions derived from it will certainly not lead to feasible covers. For the feasible minimum cover problem, if the number of lines in a partial solutions exceeds the size of a minimum cover then it is also discarded. The problem of determining a minimum cover can be solved in polynomial time, and is discussed in more detail in the next section. If either P_{include} or P_{exclude} is a feasible cover, then the solution has been found. Otherwise, these partial solutions are added to the list of existing partial solutions and the process is repeated. If the list becomes empty, then no solution exists and the algorithm halts. The property that every solution is derivable from some partial solution on the list is maintained throughout the search.

Our approach to improving the performance of incremental search is to reduce the size of the search space as much as possible without removing any solutions from the search space. Fewer partial solutions are considered during the search but no solutions to the reconfiguration

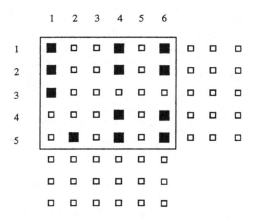

Figure 2.1: A reconfigurable array.

problem are omitted. In order to reduce the size of the search space, we make each partial solution as large as possible before the expansion step. Given a partial solution, P, let $\mathcal{S} = \{S_0, \ldots, S_k\}$ be the set of all solutions derivable from P. An assignment, A, of spare elements to faulty elements is said to be *admissible* with respect to P if all solutions in $\mathcal{S}$ are also derivable from the assignment $P \cup A$ [2]. Thus, we may use $P \cup A$ instead of P without eliminating any possible solutions from the solution space. Figure 2.2a illustrates the standard way to expand a partial solution, while Figure 2.2b illustrates our approach of enlarging the partial solution with an admissible assignment prior to expansion. The largest admissible assignment for a partial solution is said to be a *maximum admissible assignment*. In order to reduce the number of partial solutions examined as much as possible, we wish to find the maximum admissible assignment for each partial solution under consideration. Since assignments correspond to sets of lines in the case of reconfigurable arrays with spare rows and columns, admissible assignments also correspond to sets of lines. We will henceforth call these sets *admissible sets*.

---

[2]Note that an assignment is a set of ordered pairs where the ordered pair (x, y) represents the situation that spare element x is assigned to repair faulty element y.

(a)   (b)

Figure 2.2: (a) Standard expansion of partial solution. (b) Enlarging the partial solution with an admissible assignment prior to expansion.

Theorem 1 *The problem of finding a maximum admissible set for a given partial solution for the feasible minimum cover problem is NP-hard.*

Theorem 2 *The problem of finding a maximum admissible set for a given partial solution for the feasible cover problem is NP-hard.*

The proofs of these theorems are based on polynomial time Turing reductions from the feasible minimum cover problem and the feasible cover problem, respectively [50]. Although the computation of maximum admissible sets appears to be intractable, in the next two sections we provide polynomial time algorithms that find large, although not necessarily maximum, admissible sets for the feasible minimum cover problem and the feasible cover problem.

2.3   The Feasible Minimum Cover Problem

In this section we find admissible sets for the feasible minimum cover problem. These admissible sets are called *critical sets*. In Subsection 2.3.1 we introduce the mathematical tools used in the computation of critical sets and present a polynomial time algorithm to compute critical sets called the *critical set algorithm*. In Subsection 2.3.2 we describe an exhaustive search algorithm using critical sets for the feasible minimum cover problem. In Subsection 2.3.3 we provide experimental results comparing the performance of our search algorithm with an exhaustive search algorithm that does not use admissible sets.

2.3.1   Critical Sets

Let A_1 denote a given faulty array. With each partial solution, P, we associate an array, called the *residual array*, comprising the lines of A_1 that are not in the partial solution and contain faulty elements. In other words, a partial solution is represented by the faulty subarray that remains after replacing the lines in the partial solution by spare lines. The *critical set* for an array is the set of rows and columns of maximum cardinality that must be included in every minimum cover of the array. In other words, the critical set is the intersection of all minimum covers of the given faulty array. For example, the critical set for the array in Figure 2.3 consists of row 4 and column 6. This can be seen by observing that there are only two minimum covers for the faulty array, one comprising rows 1, 3, and 4 and column 6 and the other comprising row 4 and columns 1, 2, and 6. Since row 4 and column 6 appear in both minimum covers, the critical set consists of row 4 and column 6.

Consider the problem of finding a minimum cover for the faulty array A_1. Let P be a partial solution, that is, a set of lines that contain some of the faults in A_1. Let A_2 be the residual array corresponding to P, that is, the faulty subarray remaining when the lines in P are removed from A_1. Let C be the critical set for A_2. The set C is an admissible set for P since C is the set of lines that must be included in every minimum cover of A_2 and thus every minimum cover of A_1. Since the critical sets are admissible sets in the minimum cover problem, they are clearly admissible sets in the minimum feasible cover problem.

Before describing an algorithm for finding critical sets, we review some important definitions and introduce some convenient notation. An $m \times n$ faulty array can be represented by a bipartite graph $G = (X \cup Y, E)$, where X is a set of m vertices representing the rows of the array, Y is a set of n vertices representing the columns of the array, and E is a set of edges representing the faulty elements in the array. There is an edge between the vertex in X representing row i and the vertex in Y representing

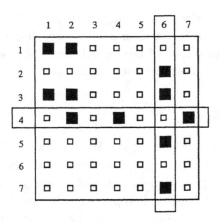

Figure 2.3: A faulty array and the corresponding critical set.

row j, if the element (i, j) in the array is faulty. An example of this representation is shown in Figure 2.4. A *matching* in a graph is a set of

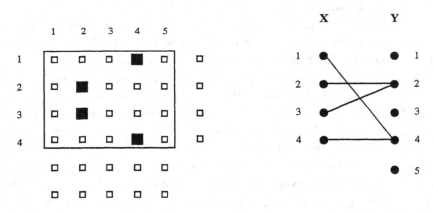

Figure 2.4: An array and its bipartite graph representation.

edges no two of which share a common vertex. The edges that belong to a given matching are called *matched edges* and those that do not belong the matching are called *unmatched edges*. Similarly, a *matched vertex* is a vertex incident to a matched edge and an *unmatched vertex* is a vertex that is not incident to any matched edge. A *maximum matching* is a matching of maximum cardinality. A *perfect matching* is a matching in which every vertex of X and every vertex of Y is matched. A *vertex cover* in a graph is a set of vertices such that every edge in the graph is incident to at least one vertex in the set. A *minimum vertex cover* is a vertex cover of minimum cardinality. Note that a vertex cover in the bipartite graph representing a faulty array corresponds to a set of rows and columns that cover all the faults in the array, that is, a cover of the array. A vertex cover composed of at most SR vertices in X and SC vertices in Y corresponds to a feasible cover of the array. Therefore, we call such a vertex cover a *feasible vertex cover*. A *feasible minimum vertex cover* is a minimum vertex cover that is also a feasible vertex cover. Thus, a solution to the feasible minimum cover problem corresponds to a feasible minimum vertex cover in the corresponding bipartite graph. Finally, a *path* in a graph is a set of edges $e_0, e_1, \ldots, e_k$ such that e_i shares a vertex with e_{i+1}, for $0 \le i \le k - 1$. Two paths are said to be *vertex disjoint* if they have no vertices in common. An *alternating path* with respect to a matching is a path in which every other edge is matched. An *augmenting path* with respect to a matching is an alternating path whose first and last edges are unmatched. The existence of an augmenting path implies that the given matching can be augmented into a new matching containing an additional matched edge.

Let A be a subset of X or a subset of Y. The *range* of A, $R(A)$, is defined to be the set of vertices that are adjacent to some vertices in A. Let $|A|$ denote the cardinality of A. The *deficiency* of A, $\delta(A)$, is defined to be $|A| - |R(A)|$. The deficiency of G from X to Y, denoted $\delta(G_{X,Y})$, is the maximum value of $\delta(A)$ over all $A \subseteq X$. Similarly, the deficiency of G from Y to X, denoted $\delta(G_{Y,X})$, is the maximum value of $\delta(A)$ over all $A \subseteq Y$. Note that $\delta(G_{X,Y})$ is not necessarily equal to $\delta(G_{Y,X})$. For example, for the bipartite graph in Figure 2.4, $\delta(G_{X,Y}) = \delta(\{1,2,3,4\}) = 4-2 = 2$ and $\delta(G_{Y,X}) = \delta(\{1,3,5\}) = 3-0 = 3$. Observe that $\delta(G_{X,Y})$ and $\delta(G_{Y,X})$ are always non-negative since the deficiency of the empty set is 0. A subset S of X is said to be a *maximum deficiency set* of X if $\delta(S) = \delta(G_{X,Y})$. Similarly, a subset S of Y is said to be a maximum deficiency set of Y if $\delta(S) = \delta(G_{Y,X})$.

The following results [53] will be used in our development of the critical set algorithm:

R1. The intersection of all maximum deficiency sets of X, denoted A_0, is also a maximum deficiency set. That is, $\delta(A_0) = \delta(G_{X,Y})$. In other words, A_0 is the smallest maximum deficiency set. Similarly, let B_0 denote the intersection of all maximum deficiency sets of Y. Then $\delta(B_0) = \delta(G_{Y,X})$. Clearly, A_0 and B_0 are unique. Therefore, the sets $R(A_0)$ and $R(B_0)$ are also unique.

R2. The maximum number of vertices in X that can be matched with the vertices in Y is $|X| - \delta(G_{X,Y})$. The maximum number of vertices in Y that can be matched with the vertices in X is $|Y| - \delta(G_{Y,X})$.

R3. The König-Egerváry Theorem [53] states that for a bipartite graph, the size of a minimum cover is equal to the size of a maximum matching. Therefore, the size of a minimum cover is equal to $|X| - \delta(G_{X,Y})$ which is also equal to $|Y| - \delta(G_{Y,X})$.

We now give a few additional results that will be used in the development of the critical set algorithm:

Lemma 1 *If $C = S \cup T$ is a minimum cover, where $S \subseteq X$ and $T \subseteq Y$, then $T = R(X - S)$.*

Proof : Since C is a cover, there is no edge from any vertex in $X - S$ to any vertex in $Y - T$. This is illustrated in Figure 2.5. Thus, $R(X - S) \subseteq T$. Suppose $R(X - S) \subset T$. That is, there is a vertex y in T that is not in $R(X - S)$. In this case, all the edges incident with y are incident with the vertices in S. Consequently, deleting y from the cover C will yield a smaller cover, which is a contradiction. $\square$

Lemma 2 *Let $C = S \cup T$ be a cover where $S \subseteq X$ and $T \subseteq Y$. Then C is a minimum cover if and only if $X - S$ is a maximum deficiency set of X.*

Proof : Let D denote $X - S$. Assume that C is a minimum cover. According to Lemma 1, $C = S \cup R(D)$. That is, the size of C is

$$|S| + |T| = |S| + |R(D)| = |X| - |D| + |R(D)|.$$

Since C is a minimum cover, it follows from R2 and R3 that

$$|X| - |D| + |R(D)| = |X| - \delta(G_{X,Y}).$$

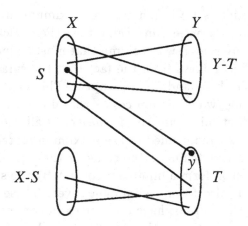

Figure 2.5: Illustration of Lemma 1.

It follows that D is a maximum deficiency set.

On the other hand, let D be a maximum deficiency set. By R2 and R3, the size of C is

$$|X| - |D| + |R(D)| = |X| - \delta(G_{X,Y}).$$

Thus, C is a minimum cover. □

Theorem 3 *A vertex $y \in Y$ is included in any minimum cover C if and only if $y \in R(A_0)$ where A_0 is the smallest maximum deficiency set of X. Similarly, a vertex $x \in X$ is included in any minimum cover C if and only if $x \in R(B_0)$ where B_0 is the smallest maximum deficiency set of Y.*

Proof : If y is included in a minimum cover C, according to Lemma 2, $y \in R(D)$ for some maximum deficiency set D. It follows that $y \in R(A_0)$ because $A_0 \subseteq D$ for any maximum deficiency set D. If $y \in R(A_0)$, then $y \in R(D)$ for any maximum deficiency set D. Again, according to Lemma 2, y is included in any minimum cover C. The second half of the theorem can be proved similarly. □

Corollary 1 *The critical set for G is equal to $R(A_0) \cup R(B_0)$.*

To find the critical set we first find a maximum matching. If the maximum matching is a perfect matching then the critical set is empty. This means that there is no row or column that must be included in every minimum cover. This follows from the fact that the smallest maximum deficiency sets of both X and Y are empty. This can also be proved directly by exhibiting two minimum covers that do not have any vertices in common. The first minimum cover consists of all the vertices in X. (All the vertices in X are included in the maximum matching. Since the size of the maximum matching is the same as the size of the minimum cover, the vertices in X form a minimum cover.) By the same reasoning, all the vertices in Y also form a minimum cover. These two covers are minimum and yet they do not have any vertex in common. This means that there is no vertex in X or Y that must be included in every minimum cover.

Let us consider the other case in which the maximum matching is not a perfect matching. That is, there are some vertices in X that are not matched with any vertex in Y or there are some vertices in Y that are not matched with any vertex in X. Let U_x be the set of unmatched vertices in X and let U_y be the set of unmatched vertices in Y. Refer to Figure 2.6, where darkened edges represent matched edges. Let A be a set consisting of all the vertices in U_x together with the vertices in X that can be reached from the vertices in U_x by following an alternating sequence of unmatched and matched edges. Let B be a set consisting of all vertices in U_y together with the vertices in Y that can be reached from the vertices in U_y by following an alternating sequence of unmatched and matched edges.

Lemma 3 *The sets A and $R(B)$ are disjoint, and the sets B and $R(A)$ are disjoint.*

Proof : Suppose that there is a vertex v in both A and $R(B)$. Then we have an alternating path between a vertex in U_x and v. This path starts with an unmatched edge and ends with a matched edge. Furthermore, there is also an alternating path between a vertex in U_y and v which starts and ends with unmatched edges. By changing the unmatched edges to matched edges and the matched edges to unmatched edges in both of these paths, we obtain a matching larger than the maximum matching, which is a contradiction.   □

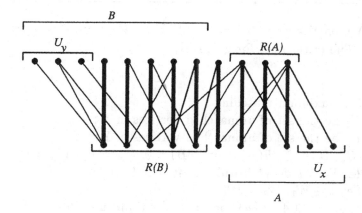

Figure 2.6: Vertex sets used in computation of critical sets.

A breadth-first search from U_x is applied in order to determine the sets A and $R(A)$ and a a breadth-first search from U_y is applied in order to determine the sets B and $R(B)$. It will be shown in Theorem 4 that the vertices in $R(A) \cup R(B)$ form the critical set for the graph G. This algorithm is shown below.

Theorem 4 *Algorithm 1 finds the critical set for G.*

Proof : First, we must show that A and B are maximum deficiency sets of X and Y, respectively. $R(A) \cup (X - A)$ is a cover for G. Since exactly one of the endpoints of each matched edge is included in the cover, the size of the cover is the same as the size of the maximum matching. Therefore, $R(A) \cup (X - A)$ is a minimum cover. By Lemma 2, A is also a maximum deficiency set of X. By the same argument, $R(B) \cup (Y - B)$ also forms a minimum cover for G. By Lemma 2, B is a maximum deficiency set of Y.

Suppose A is not the smallest maximum deficiency set of X. Let A_0 be the smallest maximum deficiency set of X. Let A_{0u} be the set of all unmatched vertices in A_0. Since $A_0 \subset A$, there exists a vertex x in $A - A_0$. Suppose that x is an unmatched vertex in X. (Refer to Figure 2.7.)

Input: A bipartite graph $G = (X \cup Y, E)$.
Output: The critical set for graph G.

begin
    Find a maximum matching $\mathcal{M}$;
    Let U_x be the set of unmatched vertices in X;
    Let U_y be the set of unmatched vertices in Y;
    $A = U_x$; $B = U_y$; $R(A) = \emptyset$; $R(B) = \emptyset$; $W_1 = U_x$; $W_2 = U_y$
    while $W_1 \neq \emptyset$ **do** (* breadth-first search *)
        Let $x \in W_1$; $W_1 = W_1 - \{x\}$;
        $R(A) = R(A) \cup \{u\}$ for $(x, u) \notin \mathcal{M}$ and $u \notin R(A)$;
        $A = A \cup \{s\}$ and $W_1 = W_1 \cup \{s\}$ for $(u, s) \in \mathcal{M}$ and $s \notin A$.
    endwhile
    while $W_2 \neq \emptyset$ **do** (* breadth-first search *)
        Let $y \in W_2$; $W_2 = W_2 - \{y\}$;
        $R(B) = R(B) \cup \{u\}$ for $(y, u) \notin \mathcal{M}$ and $u \notin R(B)$;
        $B = B \cup \{s\}$ and $W_2 = W_2 \cup \{s\}$ for $(u, s) \in \mathcal{M}$ and $s \notin B$;
    endwhile
    return$(R(A) \cup R(B))$; (* critical set is $R(A) \cup R(B)$ *)
end

Algorithm 1.

The size of a maximum matching is

$$|X| - \delta(G_{X,Y}) = |X| - (|A_0| - |R(A_0)|) = |X| - |A_{0u}|.$$

Since A_{0u} does not include all unmatched vertices, this quantity is larger than the size of a maximum matching of G, which is a contradiction. Thus, all unmatched vertices are in A_0.

Now, suppose that x is a matched vertex in X. Let (x, y) be a matched edge. Then y is in $R(A) - R(A_0)$. There is no edge from y to A_0. Otherwise, y is in $R(A_0)$. However, in this case the algorithm will never reach y from the unmatched vertices in X. Therefore, there is no such y in A. This shows that $A = A_0$. By using a similar argument on B, we show that the algorithm indeed finds the critical set for G.   □

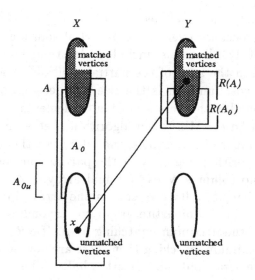

Figure 2.7: Illustration of Theorem 4.

The time taken by the critical set algorithm is the sum of the time to find the maximum matching and the time to perform breadth-first search. A maximum matching can be found in time $O(|E|\sqrt{|V|})$ [34], while breadth-first search requires time $O(|E| + |V|)$ where E is the set of edges and $V = X \cup Y$ is the set of vertices in the bipartite graph. Thus, the critical set can be found in time $O(|E|\sqrt{|V|})$.

2.3.2 An Exhaustive Search Algorithm for the Feasible Minimum Cover Problem

We now show how the concept of critical sets enables us to design an exhaustive search algorithm, called the *Min-Cover algorithm* [26], for the feasible minimum cover problem.

Let A_1 be a faulty array. We associate a cost with each partial solution which is the sum of the number of lines in the partial solution and the number of faulty elements not included in the lines in the partial solution. Notice that this cost function is non-increasing, that is:

$$\text{cost } (\{a_1, a_2, \ldots, a_k\}) \geq \text{cost } (\{a_1, a_2, \ldots, a_k, a_{k+1}\}),$$

where $a_{k+1} \neq a_\ell, \ell \leq k$ and a_{k+1} is a line that includes at least one faulty element that is not included in any a_ℓ, $\ell \leq k$. This cost function is non-increasing because every time a partial solution is expanded by adding a line to it, this line must include at least one faulty element that was not included in the lines of the partial solution. A lower bound for the value of the cost function is the size of a feasible minimum cover, and an upper bound is the number of faulty elements in the array.

We now describe the Min-Cover algorithm. Let A_1 be a faulty array, let SR be the number of spare rows, and let SC be the number of spare columns. The algorithm begins with the partial solution P_1 consisting of no rows and no columns whose residual array is A_1 itself. The cost of P_1 is the number of faulty elements in the array. First, a maximum matching is found in the bipartite graph corresponding to A_1. Denote the cardinality of the maximum matching by μ. If $SR + SC$ is less than the size of the maximum matching then there are not enough spares lines to reconfigure the array and the algorithm halts. This early termination test ensures that the algorithm will not proceed when a minimum cover cannot be realized with the available spares.

Otherwise, the critical set algorithm is applied to A_1 to find a critical set C. Since the set C is an admissible set, the lines of C are added to the empty partial solution and the cost of this new partial solution is computed. If this partial solution contains more than SR rows or SC columns then the algorithm halts and reports that no minimum feasible cover exists.

Next, the following steps are repeated until a feasible minimum cover is found or the list of partial solutions becomes empty, indicating that no feasible minimum cover exists:

1. Select a partial solution, P, of least cost.

2. Select a line, ℓ, containing the maximum number of unrepaired faulty elements.

3. Create two new partial solutions: P_{include} is $P \cup \{\ell\}$ and P_{exclude} excludes ℓ, and therefore includes all other lines that contain the faulty elements in line ℓ.

4. For each new partial solution perform the following steps:

    (a) If the partial solution is not feasible, then discard it.

(b) Otherwise, find the cardinality of the partial solution and the cardinality of a maximum matching in its residual array. Denote these quantities by s and μ', respectively.

(c) If $s + \mu' > \mu$ then a minimum cover cannot be derived from this partial solution and the partial solution is discarded.

(d) Otherwise, find the critical set for the residual array corresponding to the partial solution and add it to the partial solution.

(e) If the resulting partial solution is not feasible, then discard it.

(f) Otherwise, compute the cost of the partial solution and add the partial solution to the list of existing partial solutions.

The following example illustrates the execution of the Min-Cover algorithm.

Example: The faulty array is labeled A_1 in Figure 2.8, The faulty elements are indicated by 1's. Five spare rows ($SR = 5$) and five spare columns ($SC = 5$) are available. The tests in steps 4a and 4c of the algorithm are satisfied by every partial solution in this example and therefore are not mentioned explicitly.

The algorithm selects the partial solution $P_1 = \emptyset$ whose residual array is A_1. A maximum matching is found in the bipartite graph corresponding to A_1. The circled elements in the arrays in Figure 2.8 are the elements that correspond to edges in a maximum matching. The size of this maximum matching is 10 ($\mu = 10$) which is not more than $SR + SC$. Therefore the algorithm proceeds by finding the critical set. The critical set consists of row 5 and column 9. The new partial solution is the union of P_1 and the critical set. This partial solution, P_2, contains fewer than 5 rows and 5 columns. The cost of P_2 is found to be $2 + 18 = 20$ and P_2 is placed on the list of partial solutions. The residual array for P_2 is A_2.

Next, a row or column with the most faulty elements is selected. Here row 8 is chosen. Two new partial solutions are formed: P_3 is formed by including row 8 and P_4 is formed by excluding row 8 and thus including columns 1, 3, and 5. The residual arrays for P_3 and P_4 are denoted A_3 and A_4, respectively.

The partial solution P_3 consists of rows 5 and 8 and column 9. The critical set for the residual array A_3 consists of rows 10 and 11. The

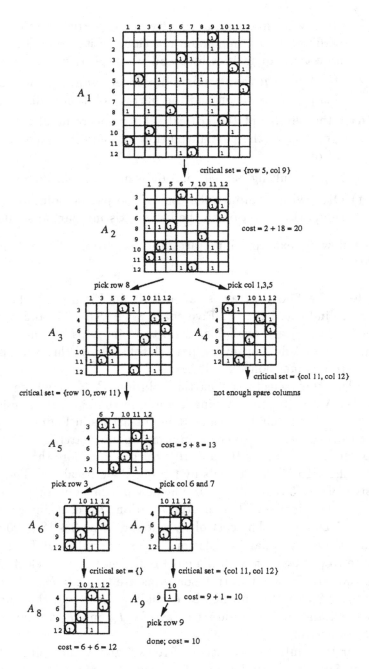

Figure 2.8: Example of the Min-Cover algorithm.

union of this set and P_3 results in a partial solution consisting of rows 5, 8, 10, and 11 and column 9. This new partial solution, labeled P_5, has no more than 5 rows and 5 columns and has cost $5 + 8 = 13$. The partial solution P_5 is added to the list of partial solutions. The residual array for P_5 is denoted A_5.

Partial solution P_4 consists of row 5 and columns 1, 3, 5, and 9. The critical set for the residual array A_4 consists of columns 11 and 12. Since the number of columns in the partial solution $P_4 \cup \{\text{column } 11, \text{column } 12\}$ is more than SC, this partial solution is discarded.

The partial solution P_5 is selected next since it is the only partial solution in the list. A row or column with the most faulty elements is selected. Here, row 3 is chosen. Again, two partial solutions are generated: P_6 is formed by including row 3 and P_7 is formed by excluding row 3, and thus including columns 6 and 7. The residual arrays for P_6 and P_7 are denoted A_6 and A_7, respectively.

Partial solution P_6 consists of rows 3, 5, 8, 10, and 11 and column 9. The critical set for the residual array A_6 is the empty set. The partial solution P_8 is the union of the critical set and P_6. This partial solution consists of rows 3, 5, 8, 10, and 11 and column 9 and has cost $6 + 6 = 12$. Since P_8 contains no more than 5 rows and 5 columns, it is added to the list of partial solutions. The residual array for P_8 is denoted by A_8.

The partial solution P_7 consists of rows 5, 8, 10, and 11 and columns 6, 7, and 9. The critical set for the residual array A_7 consists of columns 11 and 12. The partial solution P_9 is the union of the critical set and P_7. This partial solution is the set consisting of rows 5, 8, 10, and 11 and columns 6, 7, 9, 11, and 12 and has cost $9 + 1 = 10$. The corresponding residual array is denoted A_9. Since P_9 has 4 rows and 5 columns, it is added to the list of partial solutions.

The algorithm again selects a partial solution with the least cost, in this case P_9. Next, a line with greatest number of unrepaired faults is selected. Here row 9 is selected. Including row 9 in partial solution P_9 results in a feasible minimum cover for the array and the algorithm terminates. The solution consists of rows 5, 8, 9, 10, and 11 and columns 6, 7, 9, 11, and 12.

2.3.3   Experimental Results

The Min-Cover algorithm was implemented in C on an Encore Multimax computer. In addition, a simple exhaustive search algorithm was implemented on the same machine. The simple algorithm is identical to the Min-Cover algorithm except that no admissible sets are used to expand the partial solutions. The algorithms were tested on thirteen arrays constructed by Kuo and Fuchs [46] and fourteen randomly generated arrays in which the faults were uniformly distributed in clusters with varying densities.

On the arrays constructed by Kuo and Fuchs, the Min-Cover algorithm constructed approximately 15% fewer partial solutions than were constructed by the simple search algorithm. However, these problem instances were small enough that the savings in the number of partial solutions were offset by the additional computation time needed to compute the admissible sets. In fact, these problem instances were so small that the heuristic proposed in [46] only failed in one case to find a solution when one existed. Moreover, six out of the thirteen arrays had fewer spare rows and columns than the number of vertices in the minimum cover of the corresponding bipartite graph. Thus, both search algorithms immediately detected that these arrays were not repairable. Excluding these trivial cases, the reduction in the number of partial solutions was approximately 27%.

On the fourteen randomly generated arrays with clustered faults, the improvement was more dramatic. These problem instances were difficult in the sense that each array had a large number of faults and the total number of spare lines available was only slightly larger than the size of a minimum cover in the array. For one problem instance, the simple search algorithm ran for over 250 CPU seconds and constructed over 25000 partial solutions, exhausting the computer's available memory. On another problem instance the simple search algorithm ran for 240 CPU seconds and created 2622 partial solutions. In contrast, the Min-Cover algorithm ran in no more than 23 CPU seconds and created under 70 partial solutions on every problem instance. Excluding the instance in which the simple algorithm ran out of memory, the Min-Cover algorithm created an average of approximately 92% fewer partial solutions than were created by the simple search algorithm on the fourteen randomly generated arrays.

While the Min-Cover algorithm clearly created many fewer partial solutions than were created by the simple search algorithm, in several cases the difference in running times between the two algorithms was quite small. In fact, the Min-Cover algorithm even required slightly longer running time in a few cases. This phenomenon can be attributed to the overhead associated with computing the critical set for each partial solution. For small problem instances, there were few partial solutions in the search space and thus the simple algorithm could quickly enumerate all partial solutions. However, for large problem instances, the computing time incurred by the Min-Cover algorithm in finding critical sets was quite small in comparison to the time required for the simple search algorithm to construct a very large number of partial solutions. In these instances, the Min-Cover algorithm consistently had better running times than the simple search algorithm. The experimental results are summarized in Table 2.1.

2.4 The Feasible Cover Problem

In many faulty arrays, no feasible minimum covers exist. However, feasible covers for these arrays may exist. In this section we consider admissible sets for the feasible cover problem. These admissible sets are called *excess-k critical sets*. In Subsection 2.4.1 we describe excess-k critical sets and provide a polynomial time algorithm to compute these sets. In Subsection 2.4.2 we describe an exhaustive search algorithm using excess-k critical sets for the feasible cover problem and provide experimental results comparing this algorithm with other algorithms.

2.4.1 Excess-k Critical Sets

Let μ be the size of a minimum cover of a given faulty array. Our approach is to partition the feasible cover problem into the following subproblems: The problem of finding a feasible cover of size μ, the problem of finding a feasible cover of size $\mu + 1$, and so forth, up to the problem of finding a feasible cover of size $SR + SC$. Finding a solution to any one of these subproblems is equivalent to finding a solution to the feasible cover problem. We consider these subproblems in sequence until

Name	rows= cols=	Faults	SR	SC	Soln Exists?	Min-Cover Algorithm		Simple Algorithm	
						partial solutions	time (secs)	partial solutions	time (secs)
kuo1	128	5	4	4	yes	9	0.1	9	0.1
kuo2	128	15	4	4	no	0	0.1	0	0.1
kuo3	256	10	5	5	yes	15	0.1	17	0.1
kuo4	256	30	5	5	no	0	0.1	0	0.1
kuo5	512	15	5	5	yes	2	0.1	8	0.1
kuo6	512	19	10	10	yes	33	0.4	35	0.2
kuo7	512	45	10	10	no	0	0.1	0	0.1
kuo8	512	45	20	20	yes	63	2.0	73	0.4
kuo9	1024	40	20	20	yes	63	2.1	71	0.3
kuo10	1024	60	20	20	no	0	0.1	0	0.1
kuo11	1024	200	20	20	no	0	0.1	0	0.1
kuo12	1024	400	20	20	no	0	0.1	0	0.1
kuo13	1024	400	20	20	yes	17	1.5	71	1.1
RA	512	80	10	10	no	2	0.1	36	0.2
RB	512	99	17	16	yes	15	0.3	95	0.5
RC	512	107	12	12	no	2	0.1	34	0.2
RD	512	114	8	14	no	6	0.2	42	0.2
RE	512	196	22	22	no	50	2.1	330	2.0
RF	512	216	25	25	yes	7	0.5	33	0.3
RG	512	397	15	35	no	2	0.3	66	0.7
RH	1024	532	20	30	yes	3	0.5	147	1.6
RI	1024	642	12	25	no	2	0.5	40	0.6
RJ	1024	3221	50	36	no	2	4.2	102	6.9
RK	1024	966	60	60	yes	2	0.8	101	1.7
RL	1024	2715	150	200	yes	13	19.6	845	28.9
RM	1024	3996	300	200	yes	24	23.0	†	
RN	1024	3797	50	200	yes	2	7.1	2622	239.9

† Partial solutions cannot be stored in memory. Over 250 CPU secs used.

Table 2.1: Experimental results for the feasible minimum cover problem.

a solution is found or all subproblems are considered and no solution is found, in which case we can conclude that no feasible cover exists.

This approach is effective because large admissible sets can be found for the subproblems. We define the *excess-k critical set* to be the set of all lines that must be included in every cover of size $\mu + k$ or smaller. Thus, the excess-0 critical set is the same as the critical set described in the last section. Since every cover of size $\mu + k$ contains the excess-k critical set, certainly every feasible cover of size $\mu + k$ contains the excess-k critical set. Therefore, the excess-k critical sets are admissible sets for the subproblem of finding a feasible cover of size $\mu + k$.

Before showing how to compute excess-k critical sets, we describe a few of their properties that make them especially attractive as admissible sets. First, it is easily verified that the excess-k critical set is a superset of the excess-$(k + 1)$ critical set. Moreover, experimental results suggest that excess-$(k + 1)$ critical sets are only slightly smaller than excess-k critical sets and thus substantially expedite the search for a feasible cover of size $\mu+(k+1)$ if such a search is required. For example, one experiment involved a 1024×1024 array with 218 faulty elements, 15 spare rows, and 20 spare columns. After determining that $\mu = 31$, the first search using excess-0 critical sets, created only 6 partial solutions before discovering that no feasible minimum cover existed. The next search, using excess-1 critical sets, created 18 partial solutions before discovering that no feasible cover of size $\mu + 1$ existed. The search using excess-2 critical sets created 38 partial solutions, and the search using excess-3 critical sets created 70 partial solutions before discovering that no feasible covers of size $\mu+2$ and $\mu+3$ existed, respectively. Finally, the search using excess-4 critical sets found a cover of size $\mu+4$ after creating 38 partial solutions. Thus a total of 168 partial solutions were created before a solution was found. In comparison, an algorithm proposed by Hasan and Liu [26] created 21137 partial solutions before finding a feasible cover. Detailed experimental results are given in the next subsection.

Another attractive feature of using excess-k critical sets as admissible sets is that the search begins with the smallest possible feasible cover and increases the size incrementally. Therefore, if a feasible cover exists, the search will find a *minimum feasible cover*, that is, a feasible cover using the least number of lines. This is particularly advantageous in situations in which the cost of reconfiguring the array grows with the number of

spare lines used in the reconfiguration solution. Finally, the fact that
the feasible cover problem is partitioned into independent subproblems
allows the search algorithm to be easily parallelized.

We now describe an algorithm to compute excess-k critical sets.
Given a bipartite graph G and an arbitrary maximum matching $\mathcal{M}$ in G,
we introduce the following notation. Let U be the set of all unmatched
vertices with respect to $\mathcal{M}$. For any matched vertex v, let $N(v)$ be the
set of *neighbors* of v, that is, the set of all vertices adjacent to v. Let
$N_U(v)$ be the set of all unmatched neighbors of v. Also, let $C(v)$ be the
set of all vertices matched to $N(v)$. These sets are shown in Figure 2.9a,
where the edges of $\mathcal{M}$ are darkened. We define the graph $G(v)$ to be
the bipartite graph obtained by removing v and $N(v)$ and all edges incident to these vertices from G, and we define $\mathcal{M}(v)$ to be the matching
obtained from $\mathcal{M}$ when all such vertices and edges are removed. The
graph $G(v)$, corresponding to the graph G and vertex v in Figure 2.9a,
is shown in Figure 2.9b.

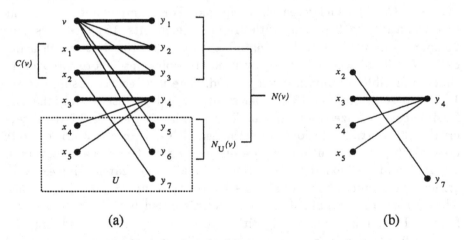

(a)   (b)

Figure 2.9: (a) The graph G and related sets. (b) The graph $G(v)$ and
matching $\mathcal{M}(v)$.

The following lemma will be used to show the correctness of our
algorithm.

Lemma 4 *Given a bipartite graph G, a maximum matching $\mathcal{M}$ for G,
and a vertex cover V' for G, V' is a minimum vertex cover iff exactly one*

endpoint of each edge of $\mathcal{M}$ is in V' and V' contains no other vertices.

Proof: Assume V' is a minimum vertex cover for G. Clearly, every edge in $\mathcal{M}$ must have at least one of its two endpoints in any vertex cover. Moreover, by the König-Egerváry Theorem $|\mathcal{M}| = |V'|$. Therefore, exactly one endpoint of each edge of $\mathcal{M}$ is included in V' and V' contains no other vertices.

Conversely, if the vertex cover V' contains exactly one endpoint of each edge of $\mathcal{M}$ and no other vertices, then $|\mathcal{M}| = |V'|$ and by the König-Egerváry Theorem, V' is a minimum vertex cover. $\square$

The following theorem tells us which vertices are in the excess-k critical set for any value of k, $0 \le k \le SR + SC - \mu$.

Theorem 5 *A vertex v is in the excess-k critical set if and only if v is matched in $\mathcal{M}$ and $|N_U(v)| + d > k$, where d is defined to be the number of vertex disjoint alternating paths in $G(v)$ with respect to $\mathcal{M}(v)$ beginning in set $C(v)$ and ending in set $U - N_U(v)$.*

Proof: Suppose that v is in the excess-k critical set. Then v is matched in $\mathcal{M}$. Otherwise, it cannot be in in the excess-k critical set since, by Lemma 4 there exists a minimum vertex cover of G containing only matched vertices. Now, assume $|N_U(v)| + d \le k$. Then there exist at most $k - |N_U(v)|$ vertex-disjoint alternating paths in $G(v)$ beginning in set $C(v)$ and ending in set $U - N_U(v)$. Assume vertex v is not included in a vertex cover of G. Then all the vertices in $N(v)$ must be included in the vertex cover. The matching $\mathcal{M}(v)$ contains $|\mathcal{M}| - (|N(v)| - |N_U(v)|)$ edges and is the largest matching in $G(v)$ that uses no vertices in $C(v)$, since a larger matching would contradict the maximality of $\mathcal{M}$. Let $\mathcal{N}$ be a maximum matching in $G(v)$. Also, let $m = |\mathcal{N}| - |\mathcal{M}(v)|$. By a well-known theorem of Edmonds [16], $G(v)$ has a set of m vertex-disjoint augmenting paths with respect to $\mathcal{M}(v)$. Each such augmenting path must have one endpoint in $C(v)$ and one endpoint in $U - N_U(v)$ or else $\mathcal{M}(v)$ is not the largest matching in $G(v)$ that uses no vertices in $C(v)$. Since $m \le k - |N_U(v)|$ by assumption, we have

$$\begin{aligned} |\mathcal{N}| &= |\mathcal{M}(v)| + m \le [(|\mathcal{M}| - (|N(v)| - |N_U(v)|)] + [k - |N_U(v)|] \\ &= |\mathcal{M}| - |N(v)| + k. \end{aligned}$$

Thus, by the König-Egerváry Theorem, the size of a minimum vertex cover of $G(v)$ is at most $|\mathcal{M}| - |N(v)| + k$. Therefore, by covering the vertices of $N(v)$ and then finding a minimum vertex cover in $G(v)$, we obtain a vertex cover of G of size at most $|\mathcal{M}| + k$. Therefore, v is not in the excess-k critical set, which leads to a contradiction.

Conversely, assume v is matched and $|N_U(v)| + d > k$. Then there exist $k - |N_U(v)| + 1$ vertex-disjoint alternating paths in $G(v)$, each with one endpoint in $C(v)$ and another endpoint in $U - N_U(v)$. Assume that v is not in a vertex cover of G. Then all vertices in $N(v)$ must be included in the vertex cover. Since $\mathcal{M}(v)$ is a matching in $G(v)$ of size $|\mathcal{M}| - (|N(v)| - |N_U(v)|)$ and each vertex-disjoint alternating path in $G(v)$ is an augmenting path, $G(v)$ has a matching of size

$$[|\mathcal{M}| - (|N(v)| - |N_U(v)|)] + [k - |N_U(v)| + 1] = |\mathcal{M}| - |N(v)| + k + 1.$$

Thus, by the König-Egerváry Theorem, the size of a minimum vertex cover of $G(v)$ is at least $|\mathcal{M}| - |N(v)| + k + 1$, and the size of a minimum vertex cover of G is at least $|\mathcal{M}| + k + 1$. Therefore, we conclude that v is in the excess-k critical set.   □

Theorem 5 tells us precisely how to compute the excess-k critical sets. For any given k, $0 \leq k \leq SR + SC - \mu$, the excess-$k$ critical set is computed as follows: First find an arbitrary maximum matching $\mathcal{M}$ in G. Next, for each vertex v matched in $\mathcal{M}$, find the sets $N(v)$, $N_U(v)$, and $C(v)$ and find the graph $G(v)$. Finally, count the number of vertex disjoint alternating paths in $G(v)$ from $C(v)$ to $U - N_U(v)$. It follows from a theorem due to Edmonds [16, 85] that this can be done by finding a single alternating path, reversing the edges on this path so that each matched edge becomes an unmatched edge and each unmatched edge becomes a matched edge, and then repeating this process until no more alternating paths can be found. Each search for an alternating path takes time $O(|E| + |V|)$ using breadth-first search [85]. From Theorem 5, at most $k + 1$ such searches are required to determine if v is in the excess-k critical set. The time complexity of this step is $O((k + 1)(|E| + |V|))$. Performing these steps for each matched vertex implies that the time required by this algorithm is $O(|E|\sqrt{|V|} + |\mathcal{M}|(k + 1)(|E| + |V|))$. The algorithm for finding an excess-k critical set is given below.

Input: A bipartite graph $G = (X \cup Y, E)$ and a non-negative integer k.
Output: The excess-k critical set for graph G.

begin
    Find a maximum matching $\mathcal{M}$;
    excess-k critical set $= \emptyset$;
    for every vertex v matched in $\mathcal{M}$ **do**
        Find the sets $N(v)$, $N_U(v)$, $C(v)$, and subgraph $G(v)$;
        alternating-paths-exist = **true**;
        number-paths-found = 0;
        while *alternating-paths-exist* **do**
            Apply breadth-first search to find an alternating path from
            $C(v)$ to $U - N_U(v)$ in $G(v)$;
            if such a path is found **then do**
                reverse matched and unmatched edges on this path;
                number-paths-found = *number-paths-found* + 1;
        endwhile
        if $|N_U(v)|+$ *number-paths-found* $> k$ **then**
            add vertex v to *excess-k critical set*;
    endfor
    return (*excess-k critical set*).
end

Algorithm 2.

2.4.2 Experimental Results

In this subsection we describe an exhaustive search algorithm for the feasible cover problem using excess-k critical sets as admissible sets. This algorithm, called the *Excess-k Cover algorithm*, is compared to the *Cover algorithm* proposed by Hasan and Liu [26].

The Excess-k Cover algorithm searches for a feasible cover of size $\mu+k$ using excess-k critical sets as the admissible sets for $0 \leq k \leq SR+SC-\mu$. The algorithm terminates as soon as a feasible cover is found or if no feasible cover is found after completing all $SR + SC - \mu + 1$ searches.

The structure of the Excess-k Cover algorithm is essentially identical to that of the Min-Cover algorithm described in Section 2.3. As in the Min-Cover algorithm, the cost assigned to each partial solution is equal to the difference between the number of faults covered by the lines of the partial solution and the number of lines in the partial solution. At each iteration of the search algorithm, a partial solution with the highest cost is selected for expansion. The line selected to expand the partial solution is the one with the largest number of unrepaired faults. These selection heuristics are the same as those used in the Cover algorithm.

A number of search algorithms for the feasible cover problem have been proposed [11, 26, 46, 54]. The algorithm proposed by Kuo and Fuchs [46] was shown to be faster than most of its predecessors. In turn, the Cover algorithm proposed by Hasan and Liu was shown to be even faster [26]. The Excess-k Cover algorithm is compared with the Cover algorithm with respect to both running time and the total number of spare lines used. If a feasible cover exists, the Excess-k Cover algorithm will find a minimum feasible cover. In contrast, neither the algorithm proposed by Kuo and Fuchs nor the Cover algorithm can guarantee finding an optimal solution.

The Excess-k Cover algorithm and the Cover algorithm were implemented in C on an Encore Multimax. The algorithms were tested on thirteen arrays constructed by Kuo and Fuchs [46] and twenty randomly generated arrays in which the faults were uniformly distributed in clusters with varying densities.

On the arrays constructed by Kuo and Fuchs, the Excess-k Cover algorithm constructed approximately 15% fewer partial solutions than the Cover algorithm. As we observed in Section 2.3, these problem instances were very small. Consequently, the savings in the number of partial solutions were offset by the additional computation time needed to compute the admissible sets. In addition, six out of the thirteen arrays had fewer spare rows and columns than the number of vertices in the minimum cover of the corresponding bipartite graph. Both search algorithms immediately determined that these arrays were unrepairable. Excluding these trivial cases, the reduction in partial solutions was approximately 28%.

The twenty randomly generated problem instances were more difficult in the sense that each array had a large number of faults and the number of spare rows and columns available was generally only slightly larger than the minimum number required. On six of the twenty problem instances, the Cover algorithm ran for over 250 CPU seconds and created more than 25000 partial solutions, exhausting the computer's available memory. In three other cases the Cover algorithm created 1846, 2192, and 21137 partial solutions before terminating. The Excess-k Cover algorithm created under 200 partial solutions and ran in under 25 CPU seconds for every problem instance. Even when excluding the cases in which the Cover algorithm ran out of memory, the Excess-k Cover algorithm created an average of 74% fewer partial solutions than were created by the Cover algorithm. Moreover, since the Excess-k Cover algorithm always finds minimum feasible covers when they exist, the feasible covers found by this algorithm were either smaller or the same size as the feasible covers found by the Cover algorithm. Excluding the cases in which the Cover algorithm ran out of memory and could not find a solution, the Excess-k Cover algorithm used approximately 10% fewer spare lines than the Cover algorithm for the twenty arrays with clustered faults.

Finally, we note that in a few cases the Excess-k Cover algorithm actually required slightly more computation time than the Cover algorithm. For very small problem instances (those with under 100 faults) this difference was typically about 0.1 CPU seconds (the smallest unit of time that the system clock reported) and could be attributed to the overhead associated with computing the excess-k critical sets for each partial solution. For larger problem instances the higher computation time resulted from the fact that the Excess-k Cover algorithm guarantees an optimal solution if any solution exists. In contrast, the Cover algorithm reports the first solution it encounters, regardless of whether the solution is optimal or not. The experimental results are summarized in Table 2.2.

2.5 Two Reconfiguration Problems

In this section we consider search algorithms for two additional reconfiguration problems in order to further illustrate the concept of admissible sets. We first consider the problem of reconfiguring a pair of arrays which share some spare rows or columns. We then consider the problem of reconfiguring redundant programmable logic arrays (RPLAs) in the

Name	rows= cols=	Faults	SR	SC	Soln Exists?	Excess-k Cover Algorithm				Cover Algorithm		
						min k	soln size	partial solns	time (secs)	soln size	partial solns	time (secs)
kuo1	128	5	4	4	yes	0	5	9	0.1	5	9	0.1
kuo2	128	15	4	4	no	0	-	0	0.1	-	0	0.1
kuo3	256	10	5	5	yes	0	9	15	0.1	9	17	0.1
kuo4	256	30	5	5	no	0	-	0	0.1	-	0	0.1
kuo5	512	15	5	5	yes	0	5	2	0.1	6	8	0.1
kuo6	512	19	10	10	yes	0	18	33	0.4	18	35	0.4
kuo7	512	45	10	10	no	0	-	0	0.1	-	0	0.1
kuo8	512	45	20	20	yes	0	38	63	1.9	38	73	1.6
kuo9	1024	40	20	20	yes	0	36	63	2.0	36	71	1.5
kuo10	1024	60	20	20	no	0	-	0	0.1	-	0	0.1
kuo11	1024	200	20	20	no	0	-	0	0.1	-	0	0.1
kuo12	1024	400	20	20	no	0	-	0	0.1	-	0	0.1
kuo13	1024	400	20	20	yes	0	36	17	1.6	37	77	2.2
R1	512	80	10	10	yes	2	18	12	0.4	20	18	0.2
R2	512	99	17	16	yes	0	30	15	0.5	33	217	1.1
R3	512	107	12	12	yes	1	19	6	0.3	24	31	0.2
R4	512	114	8	14	yes	2	22	45	0.8	22	382	1.8
R5	512	196	22	22	yes	1	43	115	5.5	44	45	1.0
R6	512	216	25	25	yes	0	35	7	0.7	45	47	0.5
R7	512	311	32	32	yes	0	64	12	2.5	†		
R8	512	397	15	35	yes	2	45	11	1.9	47	133	1.7
R9	512	574	35	45	yes	1	77	74	10.8	†		
R10	512	630	50	50	yes	0	78	9	2.5	91	32	2.2
R11	512	1069	60	60	yes	0	107	15	6.1	117	99	5.5
R12	1024	127	10	15	no	1	-	76	2.0	-	2192	9.4
R13	1024	218	15	20	yes	4	35	168	5.6	35	21137	133.2
R14	1024	927	50	60	yes	1	107	17	4.5	†		
R15	1024	1267	100	100	yes	0	120	11	6.8	150	50	6.2
R16	1024	1538	75	68	yes	0	137	15	19.2	†		
R17	1024	532	20	30	yes	0	46	3	0.5	†		
R18	1024	1084	25	20	no	2	-	6	2.9	-	308	6.6
R19	1024	642	12	25	no	2	-	6	1.7	-	1846	24.1
R20	1024	3221	50	36	no	4	-	10	23.9	†		

† Partial solutions cannot be stored in memory. Over 250 CPU secs used.

Table 2.2: Experimental results.

presence of three types of faults.

2.5.1 Reconfiguration with Shared Spares

To achieve high utilization of spares, arrays may be designed with shared spare lines. An example is shown in Figure 2.10 in which array A_1 has SR_1 of its own spare rows and array A_2 has SR_2 of its own spare rows, but the arrays share the SC spare columns. The problem of finding a feasible cover for two such arrays, which we will call the *shared spare feasible cover problem*, is defined analogously to the feasible cover problem: A *shared spare feasible cover* for the two arrays is a set of lines covering all the faults in both arrays such that at most SR_1 rows are used in A_1, at most SR_2 rows are used in A_2, and a total of at most SC columns are used by both arrays. The shared spare feasible cover problem can be easily extended to an arbitrary number of arrays sharing spare lines. Since the shared spare feasible cover problem is a generalization of the feasible cover problem, it is also NP-complete.

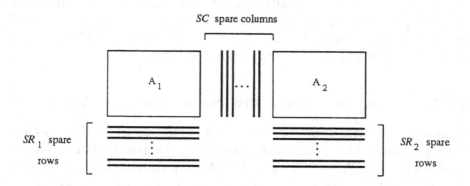

Figure 2.10: Two arrays sharing spare columns.

We describe an exhaustive search algorithm for this problem using admissible sets to reduce the size of the search space. Let μ_i be the size of a minimum cover corresponding to A_i, for $1 \leq i \leq 2$. A feasible cover for A_1 requires at least μ_1 lines but can use no more than $SR_1 + SC$ lines. Similarly, a feasible cover for A_2 requires at least μ_2 lines but can use no more than $SR_2 + SC$ lines. We therefore partition the shared spare feasible cover problem into the following subproblems: The problem of finding a feasible cover for A_1 of size μ_1 and a feasible cover for A_2 of size μ_2, the problem of finding a feasible cover for A_1 of size $\mu_1 + 1$ and a feasible cover for A_2 of size μ_2, and so forth, up to the problem of finding a feasible cover for A_1 of size $SR_1 + SC$ and a feasible cover for A_2 of size $SR_2 + SC$. We consider these subproblems in sequence until a solution is found or all subproblems are considered and no solution is found, in which case we can conclude that no shared spare feasible cover exists. An upper bound on the number of subproblems that must be considered is $(SR_1 + SC - \mu_1 + 1)(SR_2 + SC - \mu_2 + 1)$. Observe, however, that many of these subproblems can be discarded immediately since any subproblem that uses a total of more than $SR_1 + SR_2 + SC$ spare lines cannot have a shared spare feasible cover.

Such a decomposition of the shared spare feasible cover problem allows us to again use the excess-k critical sets as admissible sets. The admissible set for the subproblem of finding a feasible cover for A_1 of size $\mu_1 + i$ and a feasible cover for A_2 of size $\mu_2 + j$ is the union of the excess-i critical set for A_1 and the excess-j critical set for A_2. A number of early termination conditions exist that can reduce the number of searches performed in practice. Finally, we note that there a number of other ways to partition the shared spare feasible cover problem so that excess-k critical sets can be used as admissible sets.

2.5.2 Reconfiguration of Programmable Logic Arrays

In addition to RAMs, several other types of reconfigurable circuits have been proposed. One particularly interesting example is that of redundant programmable logic arrays (RPLAs) which have been studied in [12, 27, 72, 91, 93]. Three kinds of faults are considered in PLAs: *stuck-at* faults, *crosspoint faults*, and *bridging faults* [2]. A stuck-at fault is caused by a line being shorted to ground or V_{dd}, a bridging fault is caused by a number of adjacent or crossing lines being shorted together, and a crosspoint fault is caused by the unintended presence or absence of a transistor at the intersection of two crossing lines. Redundant PLAs with spare input, output, and product lines have been proposed by Wey [91] and by Hasan and Liu [27]. An example of an RPLA with spare lines is shown in Figure 2.11. It has been shown that the problem of reconfiguring such an RPLA with stuck-at, crosspoint, and bridging faults is NP-complete [27]. Here we show that an exhaustive search algorithm can be designed for the reconfiguration problem using admissible sets to reduce the number of partial solutions created.

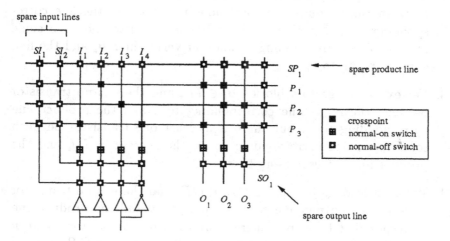

Figure 2.11: A redundant Programmable Logic Array.

The problem of reconfiguring faulty RPLAs can be formulated as a vertex cover problem in bipartite graphs with additional constraints [27]. Let $G = (X \cup Y, E)$ be a bipartite graph where the vertices in X represent the product lines of the RPLA and the vertices in $Y = I \cup O$ represent the input (I) and output (O) lines of the RPLA. Let $PB = \{PB_1, \ldots, PB_k\}$ be a partition on the set X representing the product lines of the RPLA that are bridged together. That is, PB_i is a set of vertices corresponding to a set of bridged product lines. Similarly, let $IB = \{IB_1, \ldots, IB_\ell\}$ be a partition on the set I representing bridged input lines and let $OB = \{OB_1, \ldots, OB_m\}$ be a partition on the set O representing bridged output lines. Finally, let S_P denote the number of spare product lines available, S_I denote the number of spare input lines available, S_O denote the number of spare output lines available, and $S_T = S_P + S_I + S_O$ be the total number of spare lines available. For two vertices, $x \in X$ and $y \in Y$, the edge (x, y) is in E if any of the four following conditions are satisfied:

1. There is a crosspoint fault at the intersection of the product line

corresponding to x and the input or output line corresponding to y.

2. Vertex x belongs to a block $PB_i \in PB$, and there is no transistor between the product line corresponding to x and the line corresponding to y, but there is at least one transistor between some product line corresponding to another vertex in PB_i and the line corresponding to y.

3. Vertex y belongs to a block $IB_i \in IB$, and there is no transistor between the input line corresponding to y and the product line corresponding to x, but there is at least one transistor between some input line corresponding to another vertex in IB_i and the product line corresponding to x.

4. Vertex y belongs to a block $OB_i \in OB$, and there is no transistor between the output line corresponding to y and the product line corresponding to x, but there is at least one transistor between some output line corresponding to another vertex in OB_i and the product line corresponding to x.

A *constrained vertex cover* C for graph G is defined to be a vertex cover satisfying the following six conditions:

1. $|C \cap X| \le S_P$.

2. $|C \cap I| \le S_I$.

3. $|C \cap O| \le S_O$.

4. At most one vertex of each PB_i is not included in C, for $1 \le i \le k$.

5. At most one vertex of each IB_i is not included in C, for $1 \le i \le \ell$.

6. At most one vertex of each OB_i is not included in C, for $1 \le i \le m$.

A constrained vertex cover in a bipartite graph corresponds to a solution to the RPLA reconfiguration problem [27]. By a straightforward modification of Theorem 1 we can show that the problem of finding a maximum admissible set for the constrained vertex cover problem is NP-hard. We therefore describe how large, but not necessarily maximum, admissible sets can be found for the constrained vertex cover problem.

Our approach is again to partition the problem into subproblems and find an admissible set for each subproblem. Let G be a bipartite graph with sets PB, IB, and OB as defined above, let $\mathcal{M}$ be an arbitrary maximum matching in G, and let $\mu = |\mathcal{M}|$ be the size of a minimum vertex cover of G. We partition the constrained vertex cover problem into the subproblems of finding a constrained vertex cover of size μ, a constrained vertex cover of size $\mu + 1$, and so forth up to a constrained vertex cover of size S_T. We could, of course, use the excess-k critical set as the admissible set for the subproblem of finding a constrained vertex cover of size $\mu + k$. However, the additional constraints in this problem allow us to find even larger admissible sets, called *BF-k critical sets*, that contain the excess-k critical sets as subsets.

Let $S(v)$ be the set of vertices corresponding to lines bridged with v. For example, if $v \in PB_i \subseteq X$ then $S(v) = PB_i$. If v is in no such set, then $S = \emptyset$. We define the *BF-k critical set* to be the set of all vertices v that must be included in a vertex cover of size at most $\mu + k$ such that at most one vertex in $S(v)$ is not included in the vertex cover. Given a bipartite graph G and an arbitrary maximum matching $\mathcal{M}$ in G, we again define $N(v)$ to be the set of neighbors of a matched vertex v. We define $G'(v)$ to be the subgraph of G obtained by removing v, $S(v)$, and $N(v)$ and all edges incident to these vertices from G, and we define $\mathcal{M}'(v)$ to be the matching obtained from $\mathcal{M}$ when all such vertices are removed. The following theorem tells us which vertices are in the BF-k critical set for any value of k, $0 \leq k \leq S_T$.

Theorem 6 *A vertex v is in the BF-k critical set if and only if v is matched in $\mathcal{M}$ and $|\mathcal{M}'(v)| + |N(v)| + |S(v)| + d > k + 1$, where d is defined to be the number of vertex disjoint alternating paths in $G'(v)$ with respect to $\mathcal{M}'(v)$.*

The proof of this theorem is essentially identical to the proof of Theorem 5. Moreover, this theorem tells us precisely how to compute the BF-k critical sets. The algorithm is essentially identical to Algorithm 2 except for the computation of $\mathcal{M}'(v)$ and $G'(v)$.

2.6   Summary

In this chapter we introduced the notion of admissible assignments and showed how such assignments can be used to reduce the number of partial solutions considered by exhaustive search algorithms. In the case of reconfigurable arrays with spare rows and columns, assignments correspond to sets of lines, and thus admissible assignments correspond to admissible sets. For the feasible minimum cover problem, we proposed a class of admissible sets called critical sets and gave a polynomial time algorithm for finding these sets. An exhaustive search algorithm using critical sets was given for the feasible minimum cover problem. Experimental results indicate that this algorithm examines considerably fewer partial solutions than a similar algorithm that does not use admissible sets. For the feasible cover problem, we proposed a class of admissible sets called excess-k critical sets and gave a polynomial time algorithm for finding these sets. The exhaustive search algorithm using excess-k critical sets was shown to have considerably lower running time than existing algorithms for this problem. Finally, we considered two additional applications of the admissible set approach, reconfiguration of multiple arrays using shared spares and reconfiguration of redundant programmable logic arrays.

Chapter 3

Fault Covers in Heterogeneous and General Arrays

3.1 Introduction

In the arrays discussed in Chapter 2 all elements were assumed to be identical and, thus, all rows were identical and all columns were identical. Therefore, a spare row could be used to replace any row of the array and a spare column could be used to replace any column of the array. We will subsequently refer to this as the *homogeneous array model*.

In some applications, a given row or column containing faulty elements may only be replaced by a member of a proper subset of the spare rows or spare columns, respectively. For example, consider the array shown in Figure 3.1, which contains four types of elements. In the configuration shown, the array comprises two types of rows and four types of columns. One spare row of each type is provided and one spare column of each type is provided. Clearly, a line can be replaced only by a line of the same type. This situation arises in a number of applications in which restrictions are placed on how spare lines can be used to replace faulty lines. These applications, which will be discussed later in this chapter, can be modeled by arrays containing multiple types of elements.

This chapter concerns fault cover problems in arrays containing mul-

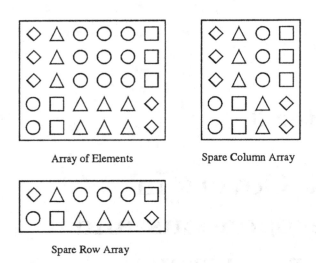

Figure 3.1: A reconfigurable array with multiple types of elements.

tiple types of elements. In Section 3.2, a fault cover model called the *heterogeneous array model* is presented. Several reconfiguration problems are considered for this model and applications of this model are discussed. In Section 3.3, a fault cover model called the *general array model* is presented. The general array model is shown to include both the homogeneous array model and the heterogeneous array model as special cases. Several reconfiguration problems are considered for this model also.

3.2   Fault Covers in Heterogeneous Arrays

In the *heterogeneous array model*, t identical copies, or replicates, of an $m \times n$ array are given, each of which may contain a pattern of faulty elements. Two additional copies of the array, called the *spare row array* and the *spare column array*, are provided as a source of spare rows and spare columns, respectively. Due to the heterogeneity of the array elements in this model, the ith row of a faulty array can be replaced only by the ith row of the spare row array and the jth column of a faulty array can be replaced only by the jth column of the spare column array. As in the homogeneous array model, a spare line in this model can be used to replace only one faulty line.

Let $A_1, \ldots, A_t$ be t $m \times n$ arrays. If L_q is a cover for A_q and L_r is a cover for A_r then L_q and L_r are said to be *disjoint* if at most one of the covers includes the ith row of its corresponding array, for $1 \leq i \leq m$, and at most one of the covers includes the jth column of its corresponding array, for $1 \leq j \leq n$. The set $\{L_1, \ldots, L_t\}$ is said to be a *feasible* set of covers for the arrays if L_ℓ is a cover for A_ℓ, for $1 \leq \ell \leq t$, and the t covers are pairwise disjoint.

From a feasible set of covers for the arrays $A_1, \ldots, A_t$, spare rows and columns can be assigned to reconfigure the arrays. Figure 3.2 shows an example of three faulty heterogeneous arrays where faulty elements are indicated by darkened squares. A feasible set of covers, indicated by arrows, is the following: Columns 1 and 2 constitute the cover for array 1, row 2 constitutes the cover for array 2, and column 4 constitutes the cover for array 3.

In this section we present results for three reconfiguration problems for the heterogeneous array model. These problems are the *feasible cover problem*, the *feasible minimum cover problem*, and the *minimum feasible cover problem*. The feasible cover problem asks whether a feasible set of covers exists for a given set of faulty arrays. The feasible minimum cover problem asks whether a feasible set of covers exists for a given set of faulty arrays such that each cover is a cover of minimum cardinality for that array. The minimum feasible cover problem seeks a feasible set of covers for a given set of faulty arrays such that the sum of the number of lines used in all the covers is minimized. In Subsections 3.2.1 and 3.2.2, we present polynomial time algorithms for the feasible cover problem and the feasible minimum cover problem, respectively. In Subsection 3.2.3 we show that the minimum feasible cover problem can be solved in polynomial time for one and two arrays ($t \leq 2$), but is NP-complete for an arbitrary number of arrays. In Subsection 3.2.4 we show that the feasible cover problem is NP-complete for a generalization of the heterogeneous array model that allows an arbitrary number of spare row arrays and spare column arrays.

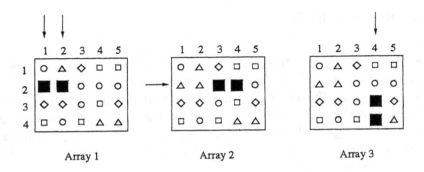

Figure 3.2: Three faulty heterogeneous arrays.

3.2.1 The Feasible Cover Problem

Given t $m \times n$ arrays, each containing zero or more faults, the *feasible cover problem* asks whether there exists a feasible set of covers for the arrays. We present a polynomial time algorithm for the feasible cover problem. Note that while we have formulated the feasible cover problem as a decision problem, the algorithm we provide is constructive and thus will produce a feasible cover if one exists. To solve the feasible cover problem, we show that the problem can be formulated as a multigraph coloring problem. This problem in turn can be reduced to the 2-satisfiability (2SAT) problem. Given a set of boolean variables $U = \{u_1, \ldots, u_k\}$, a set of clauses $\mathcal{D} = \{D_1, \ldots, D_\ell\}$ each consisting of the disjunction of at most two variables, and the conjunction of these clauses, S, the 2SAT problem asks if there exists an assignment that satisfies S. The 2SAT problem is solvable in polynomial time using any one of several known algorithms [10, 18] [1]. Note that while 2SAT is formulated as a decision problem, these algorithms will find a satisfying assignment if one exists. The multigraph is constructed as follows:

---

[1] The 2SAT algorithm in [10] is based on the following observation. Assume S contains the clauses $(x \vee y)$ and $(\neg x \vee z)$. That is, $S = (x \vee y) \wedge (\neg x \vee z) \wedge S'$. Then S is satisfiable if and only if $(y \vee z) \wedge S'$ is satisfiable. This transformation is applied repeatedly until no more literals can be eliminated, resulting in a conjunction T. It is easy to check if T is satisfiable. Moreover, T is satisfiable if and only if S is satisfiable.

Construction 1 *Given are* t $m \times n$ *replicated, heterogeneous arrays* $A_1, A_2, \ldots, A_t$. *We represent each fault, i, by a vertex v_i in the multigraph G. We associate with v_i the label $(a_i : r_i, c_i)$, where a_i is the index of the array containing the fault, r_i is the row containing the fault, and c_i is the column containing the fault. For each pair of vertices, v_i and v_j, with labels $(a_i : r_i, c_i)$ and $(a_j : r_j, c_j)$, respectively, a red edge is added between v_i and v_j if $a_i \neq a_j$ and $r_i = r_j$. Similarly, a black edge is added between v_i and v_j if $a_i \neq a_j$ and $c_i = c_j$. Let V denote the set of vertices and E the set of edges in G.*

The multigraph G may not necessarily be connected. In fact, if there exists a fault in row r and column c of an array and there are no faults in row r or column c of all other arrays, then this fault will be represented by an isolated vertex.

Next, we consider the problem of assigning the colors red and black to the vertices of such a multigraph. We say that a coloring is *feasible* if every vertex is colored, no black edge has more than one black endpoint, and no red edge has more than one red endpoint.

Theorem 7 *A feasible coloring for a multigraph resulting from Construction 1 exists if and only if there exists a feasible set of covers $\{L_1, L_2, \ldots, L_t\}$ for the arrays $A_1, A_2, \ldots, A_t$.*

Proof: In the following, $i, j, k, l \in \{1, 2, \ldots, |V|\}$. Assume that $L_1, L_2, \ldots, L_t$ are the elements of a feasible set of covers for $A_1, A_2, \ldots, A_t$. For each fault at location (r_i, c_i) of array A_{a_i}, if row r_i is contained in L_{a_i}, then we color vertex v_i red; otherwise we color the vertex black. We claim that this coloring is feasible. If not, there must exist a red edge incident to red vertices, v_i and v_j, with $r_i = r_j$, or a black edge incident to black vertices, v_k and v_ℓ, with $c_k = c_\ell$. The former case implies that row r_i is contained in two different covers, L_{a_i} and L_{a_j}. This contradicts our assumption that the covers are disjoint. The latter case implies that column c_k is contained in two different covers, L_{a_k} and L_{a_ℓ}. Again, this is a contradiction.

Next, let $C : V \to \{red, black\}$ be a feasible coloring of the multigraph. We construct a set of covers $\{L_1, L_2, \ldots, L_t\}$ as follows. For each vertex v_i that is colored red, we include row r_i in cover L_{a_i}. For each vertex v_j that is colored black, we include column c_j in cover L_{a_j}. We claim that $\{L_1, L_2, \ldots, L_t\}$ is a feasible set of covers for $A_1, A_2, \ldots, A_t$. Since each vertex represents a fault in an array, and for each vertex v_i at least one of row r_i and column c_i is included in cover L_{a_i}, it follows that $L_1, L_2, \ldots, L_t$ constitute covers for $A_1, A_2, \ldots, A_t$, respectively. Assume that $L_1, L_2, \ldots, L_t$ are not pairwise disjoint. If a row r_i is included in both L_{a_i} and L_{a_j}, then there must exist two vertices, v_i and v_j, such that $r_i = r_j$, both vertices are colored red, and the vertices share a red edge, a contradiction. Similarly, if a column c_k is included in both L_{a_k} and L_{a_ℓ}, then there must be two vertices, v_k and v_ℓ, such that $c_k = c_\ell$, both vertices are colored black, and the vertices share a black edge. Again, this is a contradiction.   □

Figure 3.3 shows the multigraph corresponding to the arrays shown in Figure 3.2. Red edges are depicted with solid lines, black edges with dashed lines. If vertices v_1, v_2, v_5 and v_6 are colored black and vertices v_3 and v_4 are colored red, then the coloring is feasible. From this solution, we generate disjoint covers for the arrays as follows: For each red vertex labeled $(a_i : r_i, c_i)$, we assign spare row r_i to A_{a_i}. For each black vertex labeled $(a_j : r_j, c_j)$, we assign spare column c_j to array A_{a_j}. That is, columns 1 and 2 are assigned to array 1; row 2 is assigned to array 2; column 4 is assigned to array 3. This is the solution shown in Figure 3.2.

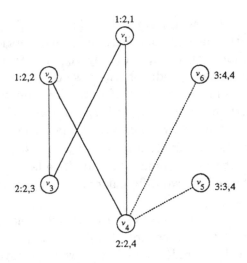

Figure 3.3: Multigraph for feasibility problem.

In order to solve the feasible cover problem, we require an algorithm to solve the multigraph coloring problem. The following construction shows that this problem can be formulated as an instance of 2SAT, solvable in polynomial time [10, 18].

Construction 2 *Given a multigraph $G = (V, E)$ with red edges and black edges, we construct a conjunction of clauses as follows: For each vertex $v_i \in V$, we introduce a boolean variable s_i. For each red edge (v_i, v_j), we include the clause $(\neg s_i \lor \neg s_j)$; for each black edge (v_k, v_ℓ), we include the clause $(s_k \lor s_\ell)$.*

Setting s_i to **true** means that vertex v_i is colored red and the corresponding fault is replaced by a row. Similarly, setting s_i to **false** means that vertex v_i is colored black and the corresponding fault is replaced by a column.

Theorem 8 *The conjunction of clauses resulting from Construction 2 is satisfiable if and only if a feasible coloring exists for the multigraph G.*

Proof: Let $C : V \rightarrow \{red, black\}$ be a feasible coloring of the multigraph G. We assign values to the boolean variables as follows: For each vertex v_i that is colored red, we assign variable s_i the value **true**. For each vertex v_j that is colored black, we assign variable s_j the value **false**. Each red edge (v_i, v_j) has at least one black endpoint, so the clause $(\neg s_i \lor \neg s_j)$ is **true**. Each black edge (v_k, v_ℓ) has at least one red endpoint, so the clause $(s_k \lor s_\ell)$ is **true**. Therefore, all of the clauses in the conjunction are **true**.

Next, let $TA : \{s_i\} \rightarrow \{$**true**, **false**$\}$ be a truth assignment satisfying the conjunction of clauses. We color the multigraph as follows: For each **true** variable, we color its corresponding vertex red. For each **false** variable, we color its corresponding vertex black. Notice that an isolated vertex will not be represented in the conjunction of clauses. For completeness, we color each isolated vertex red. Each clause of the form $(\neg s_i \lor \neg s_j)$ is **true**, so the red edge (v_i, v_j) it represents must have at least one black endpoint. Each clause of the form $(s_k \lor s_\ell)$ is **true**, so the black edge (v_k, v_ℓ) it represents must have at least one red endpoint. Therefore, the coloring is feasible. □

As an example, we give the 2SAT formulation for the set of arrays shown in Figure 3.2. Using the numbering of the vertices in Figure 3.3, the conjunction of clauses is: $(\neg s_1 \lor \neg s_3) \land (\neg s_1 \lor \neg s_4) \land (\neg s_2 \lor \neg s_3) \land (\neg s_2 \lor \neg s_4) \land (s_4 \lor s_5) \land (s_4 \lor s_6)$. A satisfying truth assignment is constructed by setting s_3 and s_4 to be **true** and setting s_1, s_2, s_5, and s_6 to be **false**.

3.2.2   The Feasible Minimum Cover Problem

The feasible minimum cover problem asks whether there exists a feasible set of covers for a given set of arrays with the additional constraint that every cover is minimum. As was discussed in Chapter 2, finding minimum covers is one way to reduce the cost of repairing the chip [11]. Again, note that while this problem is formulated as a decision problem, the algorithm given here is constructive and thus also solves the corresponding optimization problem.

As described in Chapter 2, an $m \times n$ array can be represented by a bipartite graph $G = (V = X \cup Y, E)$. Recall that vertices in X represent the rows of the array, vertices in Y represent the columns of the array, and edges in E represent faulty elements in the array. Recall also that a maximum matching in a bipartite graph can be found in time $O(|E|\sqrt{|V|})$ [34]. A minimum vertex cover in a bipartite graph can also be found in time $O(|E|\sqrt{|V|})$ by finding a minimum cut in a unit network [85] .

We now show that the feasible minimum cover problem, like the feasible cover problem, can be reduced to 2SAT. The conjunction of clauses is formed using the following construction.

Construction 3 *Let m and n be the number of rows and columns, respectively, in each of t arrays, $A_1, A_2, \ldots, A_t$. Let G_i be the bipartite graph corresponding to A_i. Let $\mathcal{M}_i$ be a maximum matching for G_i. For each row $r_i, 1 \leq r_i \leq m$, we introduce t boolean variables, $r_{i,1}, r_{i,2}, \ldots, r_{i,t}$. For each column $c_i, 1 \leq c_i \leq n$, we introduce t boolean variables, $c_{i,1}, c_{i,2}, \ldots, c_{i,t}$. The conjunction consists of four types of clauses:*

1. *For each fault in an array, we include the clause $(r_{i,k} \lor c_{j,k})$, where r_i, c_j, and A_k are the row, column, and array, respectively, that contain the fault.*

2. *Next, for each row r_i that contains a fault in one or more of the arrays, and for each unordered pair of arrays, A_k and A_ℓ, we in-*

clude the clause $(\neg r_{i,k} \lor \neg r_{i,\ell})$. *For each column c_j that contains a fault in one or more of the arrays, and for each unordered pair of arrays, A_k and A_ℓ, we include the clause* $(\neg c_{j,k} \lor \neg c_{j,\ell})$. *Hence, for each line that contains a fault in one or more of the arrays,* $\begin{pmatrix} t \\ 2 \end{pmatrix}$ *clauses are included in the conjunction.*

3. *For each fault that is represented by an edge in $\mathcal{M}_k$, we include the clause* $(\neg r_{i,k} \lor \neg c_{j,k})$, *where r_i and c_j are the row and column, respectively, that contain the fault.*

4. *Finally, for each row r_i whose representative vertex in G_k is not matched, we include the clause* $(\neg r_{i,k})$. *For each column c_j whose representative vertex in G_k is not matched, we include the clause* $(\neg c_{j,k})$.

Theorem 9 *The conjunction of clauses resulting from Construction 3 is satisfiable if and only if there exists a feasible set of minimum covers* $\{L_1, L_2, \ldots, L_t\}$ *for arrays* $A_1, A_2, \ldots, A_t$.

Proof: Assume that there exist disjoint minimum covers $L_1, L_2, \ldots, L_t$ for arrays $A_1, A_2, \ldots, A_t$. We assign truth values to variables as follows: For each spare row $r_i \in L_k$, we set $r_{i,k}$ to be **true**; for each spare column $c_j \in L_\ell$, we set $c_{j,\ell}$ to be **true**. Each fault in A_k is replaced by its row r_i or column c_j, so its corresponding clause $(r_{i,k} \lor c_{i,k})$ must be **true**. Since each spare row r_i can be assigned to at most one cover, every clause of the form $(\neg r_{i,k} \lor \neg r_{i,\ell})$ must be **true**. Since each spare column c_j can be assigned to at most one cover, every clause of the form $(\neg c_{j,k} \lor \neg c_{j,\ell})$ must be **true**. By Lemma 4, we know that, for each edge in a matching $\mathcal{M}_k$, exactly one of its endpoints must be included in a minimum cover of G_k. Therefore, every clause of the form $(\neg r_{i,k} \lor \neg c_{j,k})$ must be **true**. Finally, by Lemma 4, an unmatched vertex in a bipartite graph G_k cannot be in a minimum cover of G_k, so all 1-clauses must be **true**. Hence, using the truth assignment above, the conjunction of clauses is **true**.

Conversely, assume the conjunction is satisfiable. Then there exists a truth assignment that forces every clause to be **true**. For each **true** variable $r_{i,k}$, include spare row r_i in cover L_k. For each **true** variable $c_{j,\ell}$, include spare column c_j in cover L_ℓ. The clauses from step 1 imply that every fault is replaced. The clauses from step 2 imply that the covers are disjoint. The clauses from steps 3 and 4 imply that the covers are minimum covers. □

We omit the details of the conjunction for the example shown in Figure 3.2. We note, however, that while there exists a solution to the feasibile cover problem for this example, there does not exist a solution to the feasible minimum cover problem for the three arrays shown. Such a set could involve no more than three spare lines, but the faulty elements in the arrays cannot be replaced with fewer than four lines.

3.2.3   The Minimum Feasible Cover Problem

In this section we consider the minimum feasible cover problem for the heterogeneous array model. The *minimum feasible cover problem* is to find a feasible set of covers for a given set of arrays such that the total number of lines in all the covers is minimized. We begin by showing that this problem can be solved in polynomial time when $t \leq 2$. We then show that the problem is NP-complete for arbitrary values of t. The complexity of the minimum feasible cover problem is still open for fixed values of t greater than two.

We first observe that the minimum feasible cover problem is trivial in the case that $t = 1$, that is, there is only one array. The minimum number of lines needed to cover all the faults in the array is equal to the size of the minimum vertex cover in the corresponding bipartite graph. Moreover, the vertices in the minimum vertex cover correspond to the rows and columns in the cover of the array. Since there is only one cover, the disjointness property is vacuously true. Thus, in the case that $t = 1$, we see that the minimum feasible cover problem is equivalent to the feasible minimum cover problem.

In the case of two faulty arrays the problem becomes somewhat more difficult. Let A_1 and A_2 be two faulty $m \times n$ arrays. We wish to select the minimum number of rows and columns that contain all the faulty elements in the arrays, such that the ith rows of A_1 and A_2 are not both selected , for $1 \leq i \leq m$, and the jth columns are not both selected, for $1 \leq j \leq n$. Simply finding the minimum covers for each of the arrays independently is not sufficient, since the two minimum covers may not necessarily be disjoint. In other words, when $t = 2$, the minimum feasible cover problem is no longer equivalent to the feasible minimum cover problem.

To solve the minimum feasible cover problem for two arrays we construct a linear program. We then show that the solution to this linear programming problem corresponds exactly to a solution to the minimum feasible cover problem. The linear program can then be solved by any of a number of polynomial time algorithms [38, 89].

We begin by reviewing some definitions and results from linear programming. Let $x_1, \ldots, x_k$ be real-valued variables and let $x = [x_1, \ldots, x_k]$ be the k-dimensional vector of these variables. Also, let c be a k-dimensional vector, let r be an ℓ-dimensional vector, and let A be a $k \times \ell$ matrix. The *linear programming problem in canonical form* is to minimize $c^T x$ subject to the constraints that $x \geq 0$ and $Ax \geq r$. A vector x satisfying these constraints and achieving the minimum value is called an *optimal solution* to the linear program. The matrix A is known as the *constraint matrix*. The constraint matrix can be represented by the corresponding polytope $R(A) = \{x \mid Ax \geq r, x \geq 0\}$. There may be many solutions to a linear program, but there is always at least one optimal solution which occurs at a vertex of the polytope. Moreover, if a solution is found on a facet of a polytope then any vertex on this facet is also a solution. An integer matrix A is said to be *unimodular (UM)* if the absolute value of its determinant is 1. A matrix A is said to be *totally unimodular (TUM)* if every square non-singular submatrix of A is UM. Finally, we state two important results on totally unimodular matrices [31, 33, 60].

Lemma 5 (Hoffman and Kruskal) *If the constraint matrix of a linear program in canonical form is TUM and the vector r is integer, then all the vertices of the polytope $R(A)$ are integer.*

Lemma 6 (Heller and Tompkins) *If the constraint matrix of a linear program in canonical form is the vertex-edge incidence matrix of an undirected bipartite graph, then the constraint matrix is TUM.*

Given the two $m \times n$ arrays A_1 and A_2 for which we wish to find disjoint covers of minimum total size, we construct the two corresponding bipartite graphs $G_1 = (X_1 \cup Y_1, E_1)$ and $G_2 = (X_2 \cup Y_2, E_2)$. Again, the vertices of X_1 and Y_1 represent the rows and columns, respectively, of the array A_1. Similarly, the vertices of X_2 and Y_2 represent the rows and columns, respectively, of the array A_2. Let v_{i,X_1} and v_{j,Y_1} be the ith vertex of X_1 and the jth vertex of Y_1, respectively, and let x_1^i and y_1^j be real-valued variables representing these vertices. Similarly, let v_{i,X_2} and v_{j,Y_2} be the ith vertex of X_2 and the jth vertex of Y_2, respectively, and let x_2^i and y_2^j be real-valued variables representing these vertices, for $1 \leq i \leq m$ and $1 \leq j \leq n$. We wish to find vertex covers for these two bipartite graphs that satisfy the disjointness constraint and whose combined size is minimum. We let $x_1^i = 0$ represent the fact that vertex v_{i,X_1} is not in the vertex cover for G_1 and we let $x_1^i = 1$ represent the fact that vertex v_{i,X_1} is in the vertex cover, for $1 \leq i \leq m$. The same convention is adopted for the variables representing the vertices of Y_1, X_2, and Y_2.

Given an instance of the minimum feasible cover problem for two arrays, we can now formulate the instance as the following optimization problem: Minimize the objective function

$$\sum_{1 \leq i \leq m} x_1^i + \sum_{1 \leq j \leq n} y_1^j + \sum_{1 \leq k \leq m} x_2^k + \sum_{1 \leq \ell \leq n} y_2^\ell \qquad (3.1)$$

subject to the following constraints:

1. $x_1^i, x_2^i, y_1^j, y_2^j \in \{0, 1\}$ for $1 \leq i \leq m$, $1 \leq j \leq n$.

2. $x_1^i + y_1^j \geq 1$ for all $(i, j) \in E_1$.

3. $x_2^i + y_2^j \geq 1$ for all $(i, j) \in E_2$.

4. $x_1^i + x_2^i \leq 1$ for $1 \leq i \leq m$.

5. $y_1^j + y_2^j \leq 1$ for $1 \leq j \leq n$.

Constraint 1 ensures that every vertex is either included or excluded from a vertex cover. Constraints 2 and 3 ensure that the vertices selected form valid vertex covers for the two graphs. Constraints 4 and 5 ensure that the covers are disjoint. Finally, minimization of the objective function ensures that the disjoint vertex covers are of minimum total size.

This optimization problem can now be transformed into a linear programming problem in canonical form in which all variables are restricted to take only integer values. Although the *integer linear programming problem* is NP-complete [21] [60], we will show that the constraint matrix for this particular linear program is TUM, and thus the solution found will necessarily be integer. (In the degenerate case that a non-integer solution is found, an integer solution can be found from this solution in polynomial time.) The linear programming formulation of this optimization problem seeks to minimize the sum in Equation 3.1 subject to the constraints:

1. $x_1^i, x_2^i, y_1^j, y_2^j \geq 0$ for $1 \leq i \leq m, 1 \leq j \leq n$.

2. $x_1^i + y_1^j \geq 1$ for all $(i, j) \in E_1$.

3. $x_2^i + y_2^j \geq 1$ for all $(i, j) \in E_2$.

4. $-x_1^i - x_2^i \geq -1$ for $1 \leq i \leq m$.

5. $-y_1^j - y_2^j \geq -1$ for $1 \leq j \leq n$.

We now show that the constraint matrix corresponding to this linear program is TUM.

Theorem 10 *The constraint matrix A induced from constraints 2, 3, 4, and 5 of the above linear program, is TUM.*

Proof: To show that A is TUM, we first form a new graph $G_3 = (X_1 \cup X_2 \cup Y_1 \cup Y_2, E_1 \cup E_2 \cup E_3)$ from the bipartite graphs G_1 and G_2. The graph G_3 includes all the vertices and edges of G_1 and G_2 as well as a new edge set, E_3, containing the edges (v_{i,X_1}, v_{i,X_2}) for $1 \leq i \leq m$ and (v_{j,Y_1}, v_{j,Y_2}) for $1 \leq j \leq n$. This construction is illustrated in Figure 3.4.

It is easily verified that the graph G_3 is bipartite, since no odd cycles are introduced by our construction. The edges of E_1 are denoted by $e_1^1, e_1^2, \ldots, e_1^q$, the edges of E_2 by $e_2^1, e_2^2, \ldots, e_2^r$, and the edges of E_3 by $e_3^1, e_3^2, \ldots, e_3^s$, where $q = |E_1|$, $r = |E_2|$, and $s = |E_3|$. It is easily seen that the constraint matrix A has $q + r + s$ rows and $2(m + n)$ columns. With the $q + r + s$ rows of A, we associate the labels $e_1^1, e_1^2, \ldots, e_1^q, e_2^1, e_2^2, \ldots, e_2^r, e_3^1, e_3^2, \ldots, e_3^s$. With the $2(m + n)$ columns of A we associate the labels

$$v_{1,X_1}, v_{2,X_1}, \ldots, v_{m,X_1}, v_{1,Y_1}, v_{2,Y_1}, \ldots, v_{n,Y_1},$$

$$v_{1,X_2}, v_{2,X_2}, \ldots, v_{m,X_2}, v_{1,Y_2}, v_{2,Y_2}, \ldots, v_{n,Y_2}.$$

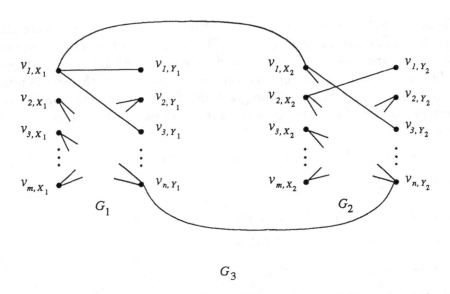

Figure 3.4: Construction of G_3 from G_1 and G_2.

The entries, $a_{i,j}$, of A are:

1. $a_{i,j} = 1$ if row i is labeled with an edge e of E_1 or E_2 and column j is labeled with a vertex v such that edge e is incident to vertex v in graph G_3. These entries represent constraints 2 and 3.

2. $a_{i,j} = -1$ if row i is labeled with an edge e of E_3 and column j is labeled with a vertex v such that edge e is incident to vertex v in graph G_3. These entries represent constraints 4 and 5.

All other entries of A are zero. Except for the fact that the rows representing constraints 4 and 5 have negative entries, this matrix is the vertex-edge incidence matrix of the bipartite graph G_3. That is, A can be obtained by multiplying some rows of the vertex-edge incidence matrix of G_3 by -1. It follows from Lemma 6 that A is TUM. □

We now show that the minimum feasible cover problem is NP-complete for arbitrary values of t. To show NP-completeness, we define the *minimum feasible cover decision problem for heterogeneous arrays* as follows: Given t replicated heterogeneous arrays and a positive integer J, does there exist a feasible set of covers for the arrays such that the total number of lines used does not exceed J?

Theorem 11 *The minimum feasible cover decision problem for hetero-geneous arrays is NP-complete.*

Proof: The problem is clearly in NP since we can guess a set of lines for each array with total size not exceeding J and verify in polynomial time that these sets constitute a feasible set of covers. To show that the problem is NP-complete, we reduce the NP-complete *vertex cover problem* [21] to the minimum feasible cover decision problem. The vertex cover problem is defined as follows: Given a graph $G = (V, E)$, and an integer $K \leq |V|$, does there exist a subset $V' \subseteq V$ with $|V'| \leq K$ such that for each edge $(u, v) \in E$, at least one of u and v belongs to V'?

Given an instance of the vertex cover problem, we construct an instance of the minimum feasible cover decision problem as follows: Let $n = |V|$ and $m = |E|$ and denote the vertices of V by $v_1, \ldots, v_n$ and the edges of E by $e_1, \ldots, e_m$. For each vertex $v_i \in V$ we construct an $m \times n(m + 1)$ faulty array, A_i. The columns of this array are divided into n *blocks*, each containing $(m + 1)$ contiguous columns. In general, the jth block contains the columns $(j - 1)(m + 1) + 1, \ldots, j(m + 1)$, for $1 \leq j \leq n$.

For each $v_i \in V$, the corresponding array, A_i, is constructed as follows: Entries in blocks other than the ith block are all non-faulty. In the ith block, the $m + 1$ elements in the jth row of the block are all faulty iff edge e_j is incident to vertex v_i in the graph G. Otherwise, the $m + 1$ elements in the jth row of the block are all non-faulty. Therefore, if edge e_j is incident upon vertices v_{i_1} and v_{i_2} then blocks i_1 of array A_{i_1} and i_2 of array A_{i_2} will each contain $(m + 1)$ faulty elements in their j^{th} rows. The jth rows of all other blocks, in all arrays, will contain only non-faulty elements. This construction can clearly be performed in polynomial time. An example of this construction is given in Figure 3.5 for a graph with 3 vertices and 3 edges.

We now claim that the graph G has a vertex cover of size K or less iff the arrays $A_1, \ldots, A_n$ have disjoint covers of total size $K(m + 1) + m$ or less. First, assume G has a vertex cover $V' \subseteq V$ such that $|V'| = \ell$ and $\ell \leq K$. For each $v_i \in V'$ all $(m + 1)$ columns in the ith block of A_i are included in the cover for A_i. These $\ell(m + 1)$ columns have distinct indices by the construction of the arrays. Since V' is a vertex cover of G, in all n arrays there exists at most one block that was not replaced by columns and still contains faulty elements in the kth row,

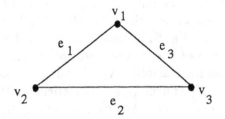

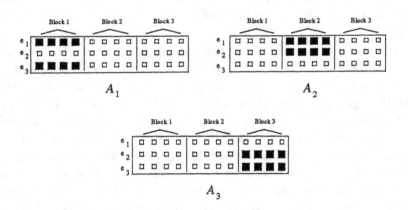

Figure 3.5: Construction of arrays from a graph.

for $1 \leq k \leq m$. Therefore, we can replace all remaining faulty elements with at most m disjoint rows. The total number of lines used is at most $\ell(m+1) + m \leq K(m+1) + m$.

Conversely, assume that there are disjoint covers for the arrays, whose union is denoted by L, using a total of $K(m+1) + m$ or fewer lines. Without loss of generality, we assume that in L either every column in a given block is in a cover or no column in the block is in a cover. If only a proper subset, S, of the columns in a block, i, are replaced, then every row containing faulty elements in block i must be replaced. Then by removing the columns in S from L, we achieve a set of lines replacing all the faults in the arrays, satisfying the disjointness property and having cardinality less than L. This implies that $|L| = \alpha(m+1) + \beta$, where α is the number of blocks repaired completely by columns and $\beta \leq m$ is the number of repaired rows. For every block, i, that is completely repaired by columns in array A_i, we include the vertex v_i in set V'. The set V' constitutes a vertex cover of G. To see this, observe that if edge e_j is incident upon vertices v_{i_1} and v_{i_2} in G, then the jth rows of A_{i_1} and A_{i_2} cannot both be in L. Thus, either block i_1 of A_{i_1} or block i_2 of A_{i_2}, or both, are completely replaced by columns. This implies that v_{i_1} or v_{i_2}, or both, are included in the set V'. So, V' is a vertex cover of size α. But, $\alpha(m+1) + \beta \leq K(m+1) + m$ and $\beta \leq m$ imply that $\alpha \leq K$. Therefore, V' is a vertex cover of G of size K or less. $\square$

We have shown that the minimum feasible cover problem for heterogeneous arrays can be solved in polynomial time when $t \leq 2$. Although the problem is NP-complete for arbitrary values of t, there may exist polynomial time algorithms for other fixed values of t.

3.2.4   The Feasible Cover Problem with Multiple Spare Arrays

The feasible cover problem with multiple spare arrays is an extension of the feasible cover problem for heterogeneous arrays discussed in Subsection 3.2.1. In this problem there may be more than one spare row array and more than one spare column array on the chip. Multiple sets of spares offer potential increases in chip yield because more faults can be successfully replaced. Of course, the increase in yield must be balanced against the increase in fabrication and materials costs accompanying the use of additional spares.

Given t $m \times n$ arrays, $A_1, \ldots, A_t$, R $m \times n$ arrays of spare rows, and C $m \times n$ arrays of spare columns, a set $\{L_1, \ldots, L_t\}$ is called a feasible set of covers for the arrays if the ith rows of at most R of the arrays are included in the covers, for $1 \leq i \leq m$, and the jth columns of at most C of the arrays are included in the covers, for $1 \leq j \leq n$. The feasible cover problem with multiple spare arrays is to determine if there exists a feasible set of covers for a given set of arrays.

An example of the multiple spare array problem, in which $R = 2$ and $C = 1$, is depicted in Figure 3.6. Unfortunately, finding such disjoint covers is much more difficult than is the original problem, in which $R = C = 1$.

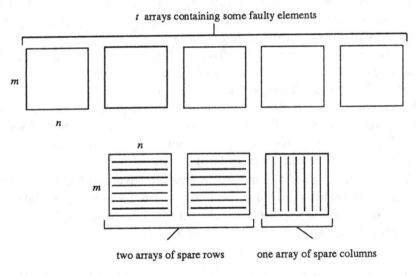

Figure 3.6: Multiple spare array problem.

Theorem 12 *The feasible cover problem with multiple spare arrays is NP-complete.*

Proof: The problem is in NP because we can select a set of lines for each array and verify in polynomial time that these sets constitute a feasible set of covers for the given arrays. Next, we must reduce a known

NP-complete problem to the feasible cover problem with multiple spare arrays. Again, our reduction is from the vertex cover problem.

Given an instance of the vertex cover problem, we construct an instance of the feasible cover problem with multiple spare arrays. Given a graph $G = (V, E)$ where $V = \{v_1, v_2, \ldots, v_n\}$, $E = \{e_1, e_2, \ldots, e_m\}$, and a positive integer $K \leq n$, we construct n $1 \times m$ arrays, $A_1, A_2, \ldots, A_n$, K $1 \times m$ arrays of spare rows, and one $1 \times m$ array of spare columns. That is, $R = K$ and $C = 1$. An entry $(1, j)$ in array A_i is faulty if and only if vertex v_i is one of the endpoints of edge e_j. This means that the number of faulty elements in any column over all the arrays is exactly 2.

We now wish to show that there exists a solution to the instance of the vertex cover problem if and only if there exists a solution to the corresponding instance of the feasible cover cover problem with multiple spare arrays. If there is a solution to the instance of the vertex cover problem, then there is a subset $V' = \{v_{i_1}, v_{i_2}, \ldots, v_{i_\ell}\}$ of V such that $\ell \leq K$ and every edge in E has at least one endpoint in V'. For each vertex v_i in V', we include row 1 in the cover for A_i. We shall use at most R spare rows, because $R = K$. Since every edge has at least one of its endpoints in V', the number of faulty elements that have not been replaced by spare rows, in any column over all the arrays, is at most 1. This means that a spare column can be used to replace each of these faulty elements.

Conversely, suppose there is a solution to the instance of the feasible cover problem with multiple spare arrays. Let $A_{i_1}, A_{i_2}, \ldots, A_{i_\ell}$, where $\ell \leq R$, be the arrays whose faulty elements are replaced by spare rows. Let $V' = \{v_{i_1}, v_{i_2}, \ldots, v_{i_\ell}\}$. Since we have only one array of spare columns, this means that the number of faulty elements not replaced by these rows in any column over all the arrays is at most 1. Recall that initially this number was 2. This means that the set V' contains at least one endpoint of each edge in E. $\qquad\square$

Although chip yield may be increased with the use of multiple sets of spares, this NP-completeness result implies that heuristic algorithms are likely to be the only viable approach to the problem. The investigation of such heuristics is a potential area for future research.

3.2.5   Applications of the Heterogeneous Array Model

The results presented above have a number of potential applications. Here we suggest two examples of such applications. First, as the density of reconfigurable arrays continues to increase, with a corresponding increase in the number of elements in the arrays, repairing a chip will require a larger number of spare lines. Often, however, an individual row or column contains only a small number of faulty elements [71]. This suggests that the approach of replacing entire rows and columns with spares will be quite inefficient since most of the elements replaced in this way will not be faulty. The number of such wasted spare elements increases as the size of the array increases and limits the probability that a faulty chip can be repaired.

One way to increase efficiency in the use of spares, and thus increase yield, is to partition the array into smaller subarrays. The spare elements are arranged such that rows and columns of individual subarrays may be replaced, independent of other subarrays, achieving the desired higher efficiency. Allocating spare lines for each subarray may be expensive. Alternatively, allowing a spare line to be used anywhere on the chip is not an attractive solution because the cost of wiring and the size of programmable decoders increases with the partitioning of the array. A compromise solution used in [29, 59] is to limit the number of subarrays to which a particular spare line may be assigned. Figure 3.7 shows how our model may be used in this manner. The array has been partitioned into 16 subarrays. The spare elements have been arranged as one array of spare rows and one array of spare columns. Row i in the spare row array can be used to replace the ith row of one subarray and column j in the spare column array can be used to replace the jth column of one subarray.

Another potential application of our model stems from recent interest in three-dimensional VLSI design [1]. Consider the situation depicted in Figure 3.8, in which eight arrays are placed between an array of spare rows and an array of spare columns. Arranging spare elements in this manner, and requiring that a spare row be used to replace only one of the rows directly below it, and that a spare column be used to replace only a column directly above it, offers one way to reduce the circuit complexity in reconfigurable three-dimensional devices.

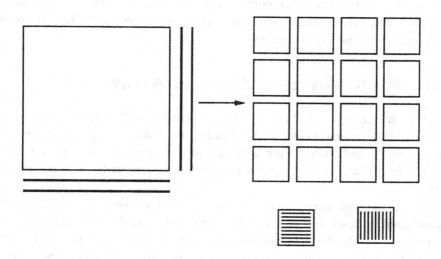

Figure 3.7: Reconfiguration with shared spares.

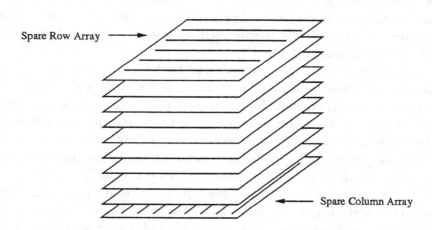

Figure 3.8: Reconfigurable stack of arrays.

Note that in both of these applications the arrays may be homogeneous. The model for heterogeneous arrays is applicable because it is assumed that the need to simplify wiring for reconfiguration imposes constraints on the use of spares.

3.3   Fault Covers in General Arrays

In this section we introduce a general reconfigurable array model that includes the homogeneous and heterogeneous array models as special cases. By investigating such a broad model, we can identify some of the features that cause array reconfiguration problems to be tractable or intractable.

In the *general array model* we are given t arrays, $A_1, \ldots, A_t$, each containing m rows and n columns of elements, some of which may be faulty. As in the heterogeneous array model, the elements might not be identical. Unlike the heterogeneous array model, however, the arrays themselves are not necessarily identical. Identical rows in the arrays are said to be of the same *type* and similarly, identical columns in the arrays are said to be of the same type. In addition to the t arrays, we are given one *spare row array* containing zero or more spare rows of each type and one *spare column array* containing zero or more spare columns of each type. A row or column containing a fault may only be replaced by a spare row or spare column, respectively, of the same type. An example of this model is shown in Figure 3.9.

Observe that the homogeneous array model is a special case of the general array model in which $t = 1$, there are SR rows in the spare row array, there are SC columns in the spare column array, and all rows and spare rows are of the same type and all columns and spare columns are of the same type. Similarly, the heterogeneous array model is a special case of the general array model in which all t arrays are identical, row i of each array is of type i and column j of each array is of type j, for $1 \leq i \leq m$ and $1 \leq j \leq n$, and the number of spare rows and spare columns of a given type is equal to 1. In the case of the heterogeneous model with multiple spare arrays, the number of spare rows of each type is R and the number of spare columns of each type is C.

As in the case of the homogeneous and heterogeneous array models, we consider three fundamental problems for the general array model:

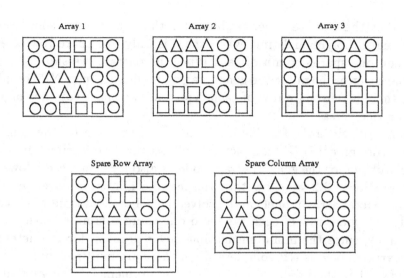

Figure 3.9: Example of the general array model.

The *feasible cover problem*, the *feasible minimum cover problem*, and the *minimum feasible cover problem*. The feasible cover problem asks whether there exists a set of covers for the arrays such that the total number of spare rows of a given type used in all the covers does not exceed the number of spare rows of that type available in the spare row array and similarly, the total number of spare columns of a given type used in all the covers does not exceed the number of spare columns of that type available in the spare column array. The feasible minimum cover problem asks whether there exists such a set of covers with the additional requirement that each cover is a minimum cover for the corresponding array. Finally, the minimum feasible cover problem is to find a feasible set of covers of minimum total size, or to determine that no feasible set of covers exists. Recall that although a feasible minimum set of covers may not exist, a minimum feasible set of covers does exist if a feasible cover exists.

Since the homogeneous array model is a special case of the general array model, and the feasible cover problem for the homogeneous array model is NP-complete, it follows that the feasible cover problem for the general array model is NP-complete. However, in Subsection 3.3.1 we show that the feasible cover problem for the general array model can be solved in polynomial time when there is only one spare row of each type and one spare column of each type. In contrast, we show that the feasible cover problem for the general array model becomes NP-complete when there is one spare row of each type and two spare columns of each type or vice versa.

Similarly, since the feasible minimum cover problem for the homogeneous array model is NP-complete, it follows that the feasible minimum cover problem for the general array model is also NP-complete. However, in Subsection 3.3.2 we show that the feasible minimum cover problem for the general array model can be solved in polynomial time when there is only one spare row of each type and one spare column of each type.

Finally, since the minimum feasible cover problem for the heterogeneous array model is NP-complete when there are multiple spare arrays, it follows that the minimum feasible cover problem for the general array model is also NP-complete. In fact, in Subsection 3.3.3 we show an even stronger result: While the minimum feasible cover problem for the heterogeneous array model is only known to be NP-complete for an arbitrary number of replicated arrays, the minimum feasible cover problem for the general array model is NP-complete for even a single array.

3.3.1   The Feasible Cover Problem

In this section we formally define the general array model and then show that the feasible cover problem for this model can be solved in polynomial time under certain constraints. We then show that the problem becomes NP-complete when these constraints are only slightly relaxed.

An instance of the general array model is a 7-tuple $(\{A_1, \ldots, A_t\}, \tau_r, \tau_c, \rho, \kappa, sr, sc)$ where

1. $A_1, \ldots, A_t$ are t $m \times n$ arrays,

2. τ_r is a non-negative integer representing the number of distinct row types,

3. τ_c is a non-negative integer representing the number of distinct column types,

4. $\rho : \{1, \ldots, t\} \times \{1, \ldots, m\} \longrightarrow \{1, \ldots, \tau_r\}$ is a surjective *row type* function mapping the rows of each array to row types,

5. $\kappa : \{1,\ldots,t\} \times \{1,\ldots,n\} \longrightarrow \{1,\ldots,\tau_c\}$ is a surjective *column type* function mapping the columns of each array to column types,

6. $sr : \{1,\ldots,\tau_r\} \longrightarrow \mathbf{Z}^+$ (where $\mathbf{Z}^+$ represents the set of non-negative integers) is a *spare row* function mapping row types to the number of spare rows of this type available,

7. $sc : \{1,\ldots,\tau_c\} \longrightarrow \mathbf{Z}^+$ is a *spare column* function mapping column types to the number of spare columns of this type available.

We define $sr_{\max} = \max_{1 \le i \le \tau_r} sr(i)$ and $sc_{\max} = \max_{1 \le j \le \tau_c} sc(j)$. The feasible cover problem for this model is to find a set of covers, $\{L_1,\ldots,L_t\}$, for arrays $A_1,\ldots,A_t$, respectively, such that the total number of rows of type i in $L_1 \cup \ldots \cup L_t$ does not exceed $sr(i)$, for $1 \le i \le \tau_r$, and the number of columns of type j in $L_1 \cup \ldots \cup L_t$ does not exceed $sr(j)$, for $1 \le j \le \tau_c$. A set of covers satisfying these constraints is called a *feasible set of covers*. In the case that $t = 1$ we say that L_1 is a *feasible cover* if $\{L_1\}$ is a feasible set of covers.

Since the homogeneous array model is a special case of the general array model, and the feasible cover problem for the homogeneous array model is NP-complete [46], it follows that the feasible cover problem for the general array model is NP-complete. However, we show that the feasible cover problem for the general array model has a polynomial time algorithm when $sr_{\max} = sc_{\max} = 1$, that is, when there is only one spare row of each type and one spare column of each type. In contrast, we then show that the feasible cover problem becomes NP-complete when $sr_{\max} = 1$ and $sc_{\max} = 2$, or $sr_{\max} = 2$ and $sc_{\max} = 1$.

We show that the feasible cover problem for the general array model when $sr_{\max} = sc_{\max} = 1$ can be solved in polynomial time by reducing any instance of the problem to an instance of the 2-satisfiability problem (2SAT), which we have noted can be solved in polynomial time as described in Subsection 3.2.1. Assume that we are given an instance of the feasible cover problem: t $m \times n$ arrays, $A_1,\ldots,A_t$, non-negative integers τ_r and τ_c, and functions ρ, κ, sr, and sc. With the ith row of array A_k we associate the boolean variable $r_{i,k}$, for $1 \le i \le m$ and $1 \le k \le t$. Similarly, with the jth column of array A_k we associate the boolean variable $c_{j,k}$, for $1 \le j \le n$ and $1 \le k \le t$. We let $r_{i,k} = \mathbf{true}$ represent the situation in which the ith row of A_k is included in the cover

L_k and let $r_{i,k} = $ **false** represent the situation in which the ith row of A_k is not included in the cover L_k. We adopt the analogous convention for the boolean variables representing columns. We now use the following construction to form an instance of 2SAT corresponding to an instance of the feasible cover problem.

Construction 4 *The boolean formula S is formed from the conjunction of the following clauses:*

1. $(r_{i,k} \lor c_{j,k})$ *for every faulty element at location (i,j) in array A_k, where $1 \le i \le m$, $1 \le j \le n$, and $1 \le k \le t$.*

2. $(\neg r_{i,k} \lor \neg r_{j,\ell})$ *for every pair of rows i of A_k and j of A_ℓ such that $\rho(k,i) = \rho(\ell,j) = \tau$ and $sr(\tau) = 1$, where $1 \le \tau \le \tau_r$, $1 \le i,j \le m$, and $1 \le k,\ell \le t$. Thus, if there are q rows of type τ, $\binom{q}{2}$ conjunctions are formed for this row type.*

3. $(\neg c_{i,k} \lor \neg c_{j,\ell})$ *for every pair of columns i of A_k and j of A_ℓ such that $\kappa(k,i) = \kappa(\ell,j) = \tau$ and $sc(\tau) = 1$, where $1 \le \tau \le \tau_c$, $1 \le i,j \le n$, and $1 \le k,\ell \le t$. Thus, if there are q columns of type τ, $\binom{q}{2}$ conjunctions are formed for this column type.*

4. $(\neg r_{i,k})$ *for every row i of A_k such that $\rho(k,i) = \tau$ and $sr(\tau) = 0$, where $1 \le \tau \le \tau_r$, $1 \le i \le m$, and $1 \le k \le t$. Thus, if there are q rows of type τ, q conjunctions are formed for this row type.*

5. $(\neg c_{j,k})$ *for every column j of A_k such that $\kappa(k,j) = \tau$ and $sr(\tau) = 0$, where $1 \le \tau \le \tau_c$, $1 \le j \le n$, and $1 \le k \le t$. Thus, if there are q columns of type τ, q conjunctions are formed for this column type.*

Since S contains only a polynomial number of clauses, it can be constructed from the arrays in polynomial time.

Theorem 13 *The boolean formula S from Construction 4 is satisfiable iff there exists a feasible set of covers $\{L_1, \ldots, L_t\}$ for the arrays $A_1, \ldots, A_t$, respectively.*

Proof: Assume S is satisfiable. Then include row i of array A_k in set L_k iff $r_{i,k} = $ **true** in the satisfying assignment and similarly, include column j of array A_k in set L_k iff $c_{j,k} = $ **true** in the satisfying assignment, for $1 \leq i \leq m$, $1 \leq j \leq n$, $1 \leq k \leq t$. First, we note that the sets $L_1, \ldots, L_t$ form covers for the arrays $A_1, \ldots, A_t$. If not, then there exist values of i, j, and k, $1 \leq i \leq m$, $1 \leq j \leq n$, $1 \leq k \leq t$, such that a faulty element in location (i, j) of array A_k is not replaced. In this case, the clause $(r_{i,k} \vee c_{j,k})$ is not satisfied, contradicting our assumption. Next, we show that these covers form a feasible set. Assume that there exist two rows, row i of A_k and row j of A_ℓ, $1 \leq i, j \leq m$ and $1 \leq k, \ell \leq t$, such that $\rho(k, i) = \rho(\ell, j) = \tau$, $1 \leq \tau \leq \tau_r$, $sr(\tau) = 1$, and the rows are included in covers L_k and L_ℓ, respectively. Then, the clause $(\neg r_{i,k} \vee \neg r_{j,\ell})$ is not satisfied, contradicting our assumption. Similarly, if two columns of the same type, τ, are included in the covers and $sc(\tau) = 1$, then a clause of the form $(\neg c_{i,k} \vee \neg c_{j,\ell})$ is not satisfied, again contradicting our assumption. If row i of array A_k is included in L_k, $\rho(k, i) = \tau$, and $sr(\tau) = 0$, $1 \leq \tau \leq \tau_r$, $1 \leq i \leq m$, $1 \leq k \leq t$, then the clause $(\neg r_{i,k})$ is not satisfied, contradicting our assumption. Similarly, if a column of type τ is included in the covers and $sc(\tau) = 0$ then a clause of the form $(\neg c_{j,k})$ is not satisfied, again contradicting our assumption.

Conversely, assume that the sets $L_1, \ldots, L_t$ are the elements of a feasible set of covers for the arrays $A_1, \ldots, A_t$. Then, for every faulty element in location (i, j) of array A_k, either row i of A_k is included in L_k or column j of A_k is included in L_k. Therefore, the clause $(r_{i,k} \vee c_{j,k})$ is satisfied. Since the covers form a feasible set, for every pair of rows i of A_k and j of A_ℓ such that $\rho(k, i) = \rho(j, \ell) = \tau$ and $sr(\tau) = 1$, at most one of $r_{i,k}$ and $r_{j,\ell}$ has the value **true**. Therefore, the clause $(\neg r_{i,k} \vee \neg r_{j,\ell})$ is satisfied. Similarly, every clause of the form $(\neg c_{i,k} \vee \neg c_{j,\ell})$ is satisfied. Finally, for every row i of A_k such that $\rho(k, i) = \tau$ and $sr(\tau) = 0$, $r_{i,k}$ has the value **false** and thus the clause $(\neg r_{i,k})$ is satisfied. Similarly, every clause of the form $(\neg c_{i,k})$ is satisfied. $\square$

We now show that the requirement that $sr_{\max} = sc_{\max} = 1$ is tight in the sense that if $sr_{\max} = 1$ and $sc_{\max} = 2$ or if $sr_{\max} = 2$ and $sc_{\max} = 1$, then the feasible cover problem for the general array model is NP-complete.

Theorem 14 *The feasible cover problem for general arrays is NP-complete even when* $sr_{\max} = 1$ *and* $sc_{\max} = 2$ *or* $sr_{\max} = 2$ *and* $sc_{\max} = 1$, *and* $t = 1$.

Proof: The problem is clearly in NP since we can guess a set of lines for the given array and verify in polynomial time that this set forms a feasible cover. Without loss of generality, we prove NP-completeness for the case that $sr_{\max} = 1$ and $sc_{\max} = 2$. By reversing the roles of rows and columns, it follows that the case in which $sr_{\max} = 2$ and $sc_{\max} = 1$ is also NP-complete.

To show that the problem is NP-complete, we reduce the NP-complete 3-satisfiability problem (3SAT) [21] to the minimum feasible cover problem. Given a set of boolean variables $U = \{u_1, \ldots, u_k\}$, a set of clauses $\mathcal{D} = \{D_1, \ldots, D_\ell\}$ each consisting of the disjunction of exactly three variables from U, and the conjunction of these clauses, S, the 3SAT problem asks if there exists an assignment that satisfies S. Given an instance of 3SAT, S, we construct a single faulty array, A, with $2k$ rows and 3ℓ columns as follows: For every variable $u_i \in U$ we associate the two rows $2i - 1$ and $2i$ of the same unique type, i, labeled with the literals u_i and $\neg u_i$, respectively. For every clause $D_j \in \mathcal{D}$ of the form $D_j = (t_j^1 \vee t_j^2 \vee t_j^3)$, where t_j^1, t_j^2, and t_j^3 represent literals, we associate the three columns $3j - 2, 3j - 1$, and $3j$ of the same unique type, j, labeled with the literals represented by t_j^1, t_j^2, and t_j^3, respectively. Array A has a fault at location (r, c) iff the label of row r is the same as the label of column c, for $1 \leq r \leq 2k$ and $1 \leq c \leq 3\ell$. We let there be exactly one spare row of each type and exactly two spare columns of each type. Thus, $sr_{\max} = 1$ and $sc_{\max} = 2$.

We now claim that the conjunction S is satisfiable iff the array A has a feasible cover. First, assume that there exists a satisfying assignment for S. From S we construct a set of rows and columns, L, as follows: For each i, $1 \leq i \leq k$, we include the row labeled u_i in L if $u_i = \textbf{true}$ in the satisfying assignment and include the row labeled $\neg u_i$ in L if $u_i = \textbf{false}$ in the satisfying assignment. We observe that for each j, $1 \leq j \leq \ell$, at least one of the three faulty elements in the three columns corresponding to D_j is replaced by a row in L. Thus, the remaining faulty elements in these three columns can be replaced by at most two columns, both of the same type. These columns are included in L. Now, L is a cover of

A, and L is a feasible cover since at most one row of each type and at most two columns of each type are included.

Conversely, assume that there exists a feasible cover, L, for the array A. Without loss of generality, we assume that exactly one row of each type is included in L. We construct a satisfying assignment for S as follows: For each i, $1 \leq i \leq k$, we set $u_i =$ **true** if the row of A labeled u_i is in L and we set $u_i =$ **false** if the row labeled $\neg u_i$ is in L. This construction results in a satisfying assignment for S. To see this, assume that there exists a clause, D_j, $1 \leq j \leq \ell$, that is not satisfied by this assignment. Then none of the three faulty elements in the three columns corresponding to the literals in D_j can be replaced by rows. However, since these faulty elements are in three distinct columns all of the same type, and $sc(j) = 2$, at least one of the faulty elements is not replaced in L. This contradicts our assumption that L is a feasible cover, and we conclude that S is satisfied by our construction. □

3.3.2 The Feasible Minimum Cover Problem

In this section we show that the feasible minimum cover problem for the general array model can be solved in polynomial time when $sr_{\max} = sc_{\max} = 1$. Assume that we are given an instance of the general array model and let μ be the minimum number of lines that contain all the faulty elements in all the arrays. We wish to find a feasible set of covers $\{L_1, \ldots, L_t\}$ for the arrays such that a total of μ lines are used, or to determine that no such cover exists.

As in the case of the feasible cover problem, we show that the feasible minimum cover problem for the general array model can be solved in polynomial time by reducing any instance of the problem to an instance of 2SAT. We again associate the boolean variable $r_{i,k}$ with the ith row of array A_k, for $1 \leq i \leq m$ and $1 \leq k \leq t$, and we associate the boolean variable $c_{j,k}$ with the jth column of array A_k, for $1 \leq j \leq n$ and $1 \leq k \leq t$. We again adopt the convention that $r_{i,k} =$ **true** iff the ith row of A_k is included in L_k and $c_{j,k} =$ **true** iff the jth column of A_k is included in L_k. Let $G_1 = (V_1 = X_1 \cup Y_1, E_1), \ldots, G_t = (V_t = X_t \cup Y_t, E_t)$ be the bipartite graphs corresponding to the arrays $A_1, \ldots, A_t$ and let $\mathcal{M}_1, \ldots, \mathcal{M}_t$ be maximum matchings for these bipartite graphs. Recall that these matchings can be found in time $O(|E|\sqrt{|V|})$ where

$E = \cup_{i=1}^{t} |E_i|$ and $V = \cup_{i=1}^{t} |V_i|$ [85]. Observe that there is a bijection, ϕ, between the set of boolean variables corresponding to the rows and columns of the arrays and the set of vertices of the bipartite graphs given by $\phi(r_{i,k}) = v_{i,X_k}$ and $\phi(c_{j,k}) = v_{j,Y_k}$ where v_{i,X_k} and v_{j,Y_k} are the ith vertex of X_k and the jth vertex of Y_k, respectively, for $1 \leq i \leq m$, $1 \leq j \leq n$, and $1 \leq k \leq t$. We now use the following construction to form an instance of 2SAT corresponding to an instance of the feasible minimum cover problem.

Construction 5 *The boolean formula S is formed from the conjunction of the following clauses:*

1. *$(r_{i,k} \vee c_{j,k})$ for every faulty element at location (i,j) in array A_k, for $1 \leq i \leq m$, $1 \leq j \leq n$, and $1 \leq k \leq t$.*

2. *$(\neg r_{i,k} \vee \neg r_{j,\ell})$ for every pair of rows i of A_k and j of A_ℓ such that $\rho(k,i) = \rho(\ell,j) = \tau$ and $sr(\tau) = 1$, where $1 \leq \tau \leq \tau_r$, $1 \leq i,j \leq m$, and $1 \leq k,\ell \leq t$. Thus, if there are q rows of type τ and one spare row of type τ, $\binom{q}{2}$ conjunctions are formed for this row type.*

3. *$(\neg c_{i,k} \vee \neg c_{j,\ell})$ for every pair of columns i of A_k and j of A_ℓ such that $\kappa(k,i) = \kappa(\ell,j) = \tau$ and $sc(\tau) = 1$, where $1 \leq \tau \leq \tau_c, 1 \leq i,j \leq n$, and $1 \leq k,\ell \leq t$. Thus, if there are q columns of type τ and one spare column of type τ, $\binom{q}{2}$ conjunctions are formed for this column type.*

4. *$(\neg r_{i,k})$ for every row i of A_k such that $\rho(k,i) = \tau$ and $sr(\tau) = 0$, where $1 \leq \tau \leq \tau_r$, $1 \leq i \leq m$, and $1 \leq k \leq t$. Thus, if there are q rows of type τ and no spare rows of type τ, q conjunctions are formed for this row type.*

5. *$(\neg c_{j,k})$ for every column j of A_k such that $\kappa(k,j) = \tau$ and $sr(\tau) = 0$, where $1 \leq \tau \leq \tau_c$, $1 \leq j \leq n$, and $1 \leq k \leq t$. Thus, if there are q columns of type τ and no spare columns of type τ, q conjunctions are formed for this column type.*

6. *$(\neg r_{i,k}) \vee \neg c_{j,k})$ for every edge $(i,j) \in M_k$, for $1 \leq i \leq m$, $1 \leq j \leq n$, and $1 \leq k \leq t$.*

7. $(\neg r_{i,k})$ for every $r_{i,k}$ such that $\phi(r_{i,k})$ is unmatched in $\mathcal{M}_k$, for $1 \leq i \leq m$ and $1 \leq k \leq t$.

8. $(\neg c_{j,k})$ for every $c_{j,k}$ such that $\phi(c_{j,k})$ is unmatched in $\mathcal{M}_k$, for $1 \leq j \leq n$ and $1 \leq k \leq t$.

Since S contains only a polynomial number of clauses, it can be constructed from the arrays in polynomial time.

Theorem 15 *The boolean formula S from Construction 5 is satisfiable iff there exists a feasible set of covers $\{L_1, \ldots, L_t\}$ for the arrays $A_1, \ldots, A_t$, respectively, such that $|L_1| + \ldots + |L_t| = \mu$.*

Proof: Assume S is satisfiable. Then include row i of array A_k in set L_k iff $r_{i,k} = \textbf{true}$ in the satisfying assignment and similarly, include column j of array A_k in set L_k iff $c_{j,k} = \textbf{true}$ in the satisfying assignment, for $1 \leq i \leq m$, $1 \leq j \leq n$, and $1 \leq k \leq t$. We first note that the sets $L_1, \ldots, L_t$ form covers for the arrays $A_1, \ldots, A_t$, respectively. If not, then there exist values of i, j, and k, $1 \leq i \leq m$, $1 \leq j \leq n$, and $1 \leq k \leq t$, such that a faulty element in location (i, j) of array A_k is not replaced. In this case, the clause $(r_{i,k} \vee c_{j,k})$ is not satisfied, contradicting our assumption. We next show that these covers form a feasible set. Assume that there exist two rows, row i of A_k and row j of A_ℓ, $1 \leq i, j \leq m$ and $1 \leq k, \ell \leq t$, such that $\rho(k, i) = \rho(\ell, j) = \tau$, $1 \leq \tau \leq \tau_r$, $sr(\tau) = 1$, and the rows are included in covers L_k and L_ℓ, respectively. Then, the clause $(\neg r_{i,k} \vee \neg r_{j,\ell})$ is not satisfied, contradicting our assumption. Similarly, if two columns of the same type, τ, are included in the covers and $sc(\tau) = 1$, then a clause of the form $(\neg c_{i,k} \vee \neg c_{j,\ell})$ is not satisfied, again contradicting our assumption. If row i of array A_k is included in L_k, $\rho(k, i) = \tau$ and $sr(\tau) = 0$, $1 \leq \tau \leq \tau_r$, $1 \leq i \leq m$, $1 \leq k \leq t$, then the clause $(\neg r_{i,k})$ is not satisfied, contradicting our assumption. Similarly, if a column of type τ is included in the covers and $sc(\tau) = 0$ then a clause of the form $(\neg c_{j,k})$ is not satisfied, again contradicting our assumption. Finally, we note that the total size of these covers is exactly μ. By definition, $\mu = |\mathcal{M}_1| + \ldots + |\mathcal{M}_t|$. Assume that there exists a set L_k, $1 \leq k \leq t$, such that $|L_k| > |\mathcal{M}_k|$. Then, by the König-Egerváry Theorem, those vertices of V_k corresponding to the lines in L_k form a vertex cover of G_k that is not a minimum cover. Then, by Lemma 4, either two vertices of

V_k adjacent in $\mathcal{M}_k$ are replaced or a vertex of V_k unmatched in $\mathcal{M}_k$ is replaced in this vertex cover. In the first case, the clause $(\neg r_{i,k} \vee \neg c_{j,k})$, where $(i,j) \in \mathcal{M}_k$ and $\phi(r_{i,k})$ and $\phi(c_{j,k})$ are replaced, is not satisfied. In the second case, the clause $(\neg r_{i,k})$ or the clause $(\neg c_{j,k})$, where $\phi(r_{i,k})$ and $\phi(c_{j,k})$ are unmatched vertices in $\mathcal{M}_k$, is not satisfied. In both cases the assumption that S is satisfiable is contradicted.

Conversely, assume that the sets $L_1, \ldots, L_t$ form a feasible set of covers of total size μ for the arrays $A_1, \ldots, A_t$. Then, for every faulty element in location (i,j) of array A_k, $1 \le i \le m$, $1 \le j \le n$, and $1 \le k \le t$, either row i of A_k is included in L_k or column j of A_k is included in L_k. Therefore, the clause $(r_{i,k} \vee c_{j,k})$ is satisfied. Since the covers form a feasible set, for every pair of rows i of A_k and j of A_ℓ such that $\rho(k,i) = \rho(j,\ell) = \tau$ and $sr(\tau) = 1$, at most one of $r_{i,k}$ and $r_{j,\ell}$ has the value **true**. Therefore, the clause $(\neg r_{i,k} \vee \neg r_{j,\ell})$ is satisfied. Similarly, every clause of the form $(\neg c_{i,k} \vee \neg c_{j,\ell})$ is satisfied. For every row i of A_k such that $\rho(k,i) = \tau$ and $sr(\tau) = 0$, $r_{i,k}$ has the value **false** and thus the clause $(\neg r_{i,k})$ is satisfied. Similarly, every clause of the form $(\neg c_{i,k})$ is satisfied. Finally, since each cover corresponds to a minimum cover in the bipartite graph, Lemma 4 implies that the clauses of the form $(\neg r_{i,k} \vee \neg c_{j,k})$, $(\neg r_{i,k})$, and $(\neg c_{j,k})$ are satisfied.   □

3.3.3   The Minimum Feasible Cover Problem

In this section we consider the minimum feasible cover problem for the general array model. Since the heterogeneous array model is a special case of the general array model and the minimum feasible cover problem for the heterogeneous array model was shown to be NP-complete in Subsection 3.2.3, it follows that the minimum feasible cover problem for the general array model is NP-complete. However, while the minimum feasible cover problem for the heterogeneous array model was shown to have a polynomial time algorithm when there are at most two replicated arrays, we show that the minimum feasible cover problem for the general array model is NP-complete even when $t = 1$.

To show NP-completeness, we define the *minimum feasible cover decision problem for general arrays* as follows: Given t $m \times n$ arrays, non-negative integers τ_r and τ_c, functions ρ, κ, sr, and sc, and a positive integer K, does there exist a feasible set of covers $\{L_1, \ldots, L_t\}$ for the

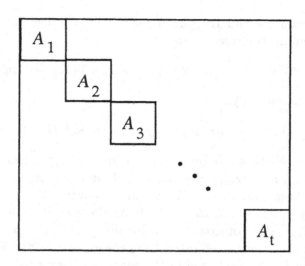

Figure 3.10: Construction of array A from arrays $A_1, \ldots, A_t$.

arrays $A_1, \ldots, A_t$, respectively, whose total size does not exceed K?

Theorem 16 *The minimum feasible cover decision problem for general arrays is NP-complete even if $t = 1$ and $sr_{\max} = sc_{\max} = 1$.*

Proof: The problem is clearly in NP since we can guess a set of lines of total size K and verify in polynomial time that they form a feasible set of covers. To show that the problem is NP-complete, we reduce the minimum feasible cover decision problem for heterogeneous arrays to the minimum feasible cover decision problem for general arrays.

Given an instance of the minimum feasible cover decision problem for heterogeneous arrays in the form of t $m \times n$ arrays, $A_1, \ldots, A_t$, a single spare row array, a single spare column array, and a positive integer J, we construct a single $tm \times tn$ array, A, as follows: The array A is partitioned into t^2 $m \times n$ subarrays such that subarray $A(k, \ell)$ is the intersection of rows $(k-1)m+1, \ldots, km$ and columns $(\ell-1)n+1, \ldots, \ell n$, for $1 \leq k, \ell \leq t$. Subarray $A(k, \ell)$ contains only non-faulty elements if $k \neq \ell$ and $A(k, k) = A_k$, for $1 \leq k, \ell \leq t$. This construction is illustrated in Figure 3.10.

To complete our reduction, we now describe how the row type function ρ, the column type function κ, the spare row function sr, the spare column function sc, and the parameter K are chosen. We assign row types and column types as follows:

$$\rho(1, i) = \rho(1, m + i) = \rho(1, 2m + i) = \ldots = \rho(1, (t - 1)m + i) = i$$

for $1 \leq i \leq m$. Similarly,

$$\kappa(1, j) = \kappa(1, n + j) = \kappa(1, 2n + j) = \ldots = \kappa(1, (t - 1)n + j) = j$$

for $1 \leq j \leq n$. Next, we define $sr(i) = 1$, for $1 \leq i \leq m$, and $sc(j) = 1$, for $1 \leq j \leq n$. Thus, $sr_{\max} = sc_{\max} = 1$. Finally, we let $K = J$.

We now claim that there exists a feasible cover L for the constructed general array A such that $|L| \leq J$ iff the there exists a feasible set of covers with total size not exceeding J for the given heterogeneous arrays. Assume that the constructed instance of the general array problem has a feasible cover, L , of size J or less. We construct covers $L_1, \ldots, L_t$ for the heterogeneous arrays $A_1, \ldots, A_t$ as follows: For every row $r \in L$, include row $r \bmod m$ of array A_k in L_k where $k = \lceil \frac{r}{m} \rceil$. For every column $c' \in L$, include row $c \bmod n$ of array A_k in L_k where $k = \lceil \frac{c}{n} \rceil$. Clearly,

$$\sum_{k=1}^{t} |L_i| = |L| \leq J.$$

Therefore, we must only show that $\{L_1, \ldots, L_t\}$ is a feasible set of covers for the arrays. First, assume that there exist i, j, and k, $1 \leq i \leq m$, $1 \leq j \leq n$, and $1 \leq k \leq t$, such that location (i, j) of array A_k contains a faulty element and neither row i nor column j are included in L_k. Then, by the construction of A, there exists a faulty element at location $((k - 1)m + i, (k - 1)n + j)$ of array A that is not replaced by a row or column in L, contradicting our assumption that L is a cover for A. Next, assume that there exist two rows, row i of array A_k and row i of array A_ℓ, that are included in covers L_k and L_ℓ, respectively. Then rows $(k - 1)m + i$ and $(\ell - 1)m + i$ of array A are included in L. However, by the construction of A these rows are of the same type, contradicting our assumption that L is a feasible cover. Similarly, there cannot exist two columns with the same indices in $L_1, \ldots, L_t$. Therefore, $\{L_1, \ldots, L_t\}$ is a feasible set of covers of total size J or less.

Conversely, assume that $A_1, \ldots, A_t$ have a feasible set of covers $\{L_1, \ldots, L_t\}$ of total size J or less. We construct a set L as follows:

For each row i of A_k included in L_k, include row $(k-1)m + i$ of A in set L, for $1 \leq i \leq m$ and $1 \leq k \leq t$. Similarly, for each column j included in L_k, include column $(k-1)n + j$ of A in set L, for $1 \leq j \leq n$ and $1 \leq k \leq t$. Again,

$$|L| = \sum_{k=1}^{t} |L_i| \leq J$$

and thus we need only show that L is a feasible cover for A. First, if L is not a cover then there must exist r and c, $1 \leq r \leq tm$ and $1 \leq c \leq tn$, such that location (r, c) of array A contains a faulty element but neither row r nor column c are contained in L. By the construction of A, this implies that location $(r \bmod m, c \bmod n)$ of A_k, where $k = \lceil \frac{r}{m} \rceil = \lceil \frac{c}{n} \rceil$, contains a faulty element which is not replaced by a line in L_k, contradicting our assumption that L_k is a cover for A_k. Next, assume there exist two distinct rows of A, row u and row v, $1 \leq u, v \leq tm$, such that rows u and v are of the same type and are both included in L. Then $u = (k-1)m + r$ and $v = (\ell-1)m + r$ for some $1 \leq k, \ell \leq t$ and $k \neq \ell$ and for some $1 \leq r \leq m$. Then L_k and L_ℓ contain row r of A_k and row r of A_ℓ, respectively, contradicting the assumption that L_k and L_ℓ are elements of a feasible set of covers. Similarly, L cannot contain two columns of the same type, and thus L is a feasible cover. $\square$

3.4 Summary

In this chapter we have investigated fault cover problems in arrays of heterogeneous elements. We first considered the heterogeneous array model in which a set of replicated arrays share a single spare row array and a single spare column array with the constraint that a row or column in a faulty array can only be replaced by a spare row or column, respectively, with the same index. We considered the feasible cover, feasible minimum cover, and minimum feasible cover problems for this model. Polynomial time algorithms were given for the feasible cover and feasible minimum cover problems. Next, we gave an algorithm for the minimum feasible cover problem for two replicated arrays and showed that the minimum feasible cover problem is NP-complete for an arbitrary number of replicated arrays. Finally, we showed how this model can apply to reconfiguration problems in other reconfigurable chip designs.

We next proposed the general array model, a model containing both the homogeneous and heterogeneous array models as special cases. We again considered the feasible cover, feasible minimum cover, and minimum feasible cover problems for this model. We gave a polynomial time algorithm for the feasible cover problem when there is at most one spare row of each type and one spare column of each type. The feasible cover problem was shown to be NP-complete when there is at most one spare row of each type and two spare columns of each type or vice versa. While the feasible minimum cover problem is NP-complete for this model, we gave a polynomial time algorithm for the case that there is at most one spare row of each type and one spare column of each type. Finally, we showed that the minimum feasible cover problem is NP-complete for this model even under very tight restrictions. By studying various restrictions to this broad model, we have identified some of the features that make reconfiguration problems tractable or intractable. Other restrictions of the general array model could also be considered, and this is an interesting topic for future research.

Chapter 4

General Formulation of Fault Covering Problems

4.1 Introduction

The relationship between faulty elements and spare elements varies for different classes of reconfigurable chips. In general, this relationship could be very complicated. For example, several spare elements may be required to replace a single faulty element or a spare element may be used to replace several faulty elements in a covering assignment. We begin with an example of a reconfigurable chip illustrating some of the possible relationships between faulty elements and spare elements.

Example 4.1: Consider the chip shown in Figure 4.1 which contains four faulty elements, u, v, w, x, and three spare elements, a, b, c. The following constraints are imposed:

1. Element u can be replaced by either a or b, v can be replaced by both b and c, w can be replaced either by both a and b or by c, and x can be replaced by b.

2. Spare a can replace only one faulty element and each of b and c can replace two faulty elements at the same time.

3. Element x must be replaced and at least two of u, v, w must be replaced.

4. At most one of a and b can be used to replace faulty elements.

In this example, one possible covering assignment allocates spare element b to both faulty elements v and x and allocates spare element c to both faulty elements v and w. Another possible covering assignment allocates spare element b to both faulty elements u and x and allocates spare element c to faulty element w.

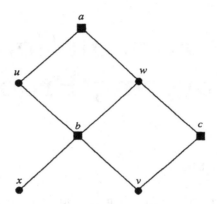

Figure 4.1: The chip of Example 4.1.

In Section 4.2, we present a general formulation which captures a large class of the possible relationships between spare elements and faulty elements in reconfigurable chips. A general formulation has two attractive features. First, it provides a framework for constructing general algorithms for the solution of a large class of fault covering problems. Second, it enables us to characterize the complexity of a large class of fault covering problems. In Section 4.3 we provide examples demonstrating how several well-known fault covering problems can be represented in the general formulation. In Section 4.4 we show how a fault covering problem in the general formulation can be transformed into an integer linear programming problem. Thus, the general formulation combined with the transformation to the integer linear programming problem provides a general and uniform approach to the solution of a large class of fault covering problems. Although the integer linear programming problem is NP-complete [21], it is an important optimization problem that has been studied extensively. Moreover, a number of software packages exist that can be used to find either an exact or an approximate solution to the problem in a reasonable amount of computation time. We provide experimental results for three problems, demonstrating the effectiveness of this approach. Finally, in Section 4.5 we characterize the complexity of the fault covering problems for sixteen subcases of the general formulation. Polynomial time algorithms are given for some subcases whereas the rest are shown to be NP-complete.

4.2   A General Formulation

We have observed that the relationship between faulty elements and spare elements could be very complicated. In this section we propose a general formulation that captures a large class of the possible relationships between faulty elements and spare elements. In particular, this formulation accounts for the following possible constraints:

1. Each faulty element can be replaced by any one of several sets of spare elements.

2. Each spare element can be used in the replacement of a number of faulty elements.

3. The faulty elements are partitioned into sets and each set is assigned a threshold value with the constraint that the number of elements in each set that are replaced must be no less than the threshold value.

4. The spare elements are partitioned into sets and each set is assigned a threshold value with the constraint that the number of spares in a set used to replace faulty elements cannot exceed the threshold value.

Constraint 1 arises when a faulty element comprises a set of components and multiple copies of these components are available as spares. Constraint 2 arises when a spare element comprises one or more copies of a regular element. For example, in the homogeneous array model discussed in Chapter 2, a spare row in an $m \times n$ array comprises n elements and thus each spare element can be used to replace n elements in the array. Constraint 3 arises in redundant systems that function if and only if at least some number k out of the available n elements are functioning. Finally, constraint 4 arises when the spare elements share some resources, such as wires and switches, that allow no more than a fixed number of the spares to be used.

The formulation we propose uses a generalized bipartite graph to represent the relationship between faulty elements and spare elements. Let $G = (X \cup Y, E, W, \delta, P_X, P_Y)$ be a generalized bipartite graph in which X and Y are two disjoint subsets of vertices, E is a set of edges with one endpoint in X and one endpoint in Y, W is a function mapping each vertex in X to an expression in the form *sum-of-products* in which the variables are vertices in Y, δ is a function mapping each vertex in Y to a positive integer, and P_X and P_Y are partitions on the vertices in X and Y, respectively.

Each vertex $x \in X$ represents a faulty element in the chip, and each vertex $y \in Y$ represents a spare element in the chip. For each vertex $x \in X, W(x) = y_{1,1}y_{1,2}\cdots y_{1,i_1} + y_{2,1}y_{2,2}\cdots y_{2,i_2} + \cdots + y_{m,1}y_{m,2}\cdots y_{m,i_m}$ means that the faulty element x can be repaired by the spare elements $y_{1,1}, y_{1,2}, \ldots, y_{1,i_1}$ or by the spare elements $y_{2,1}, y_{2,2}, \ldots, y_{2,i_2}$, etc. There is an edge between vertex x in X and vertex y in Y if y appears in $W(x)$. Given a subset of edges $A \subseteq E$, the *evaluation of $W(x)$ with respect to A*, denoted $W_A(x)$, is computed as follows: Each vertex y in $W(x)$ is treated as a boolean variable. For each variable y in $W(x)$, if $(x, y) \in A$ then the value of the boolean variable y is set to be **true**, otherwise, the value of the boolean variable y is set to be **false**. The value of $W_A(x)$ is the value of $W(x)$ based on this assignment of boolean values to the variables. A vertex x is called a **true** vertex with respect to A if $W_A(x)$ is **true**, and a **false** vertex otherwise. The expression $W(x)$ captures the situation in which one of several combinations of spare elements may be used to repair a faulty element. Each combination is represented by a product term in $W(x)$. By setting the value of variable y in $W(x)$ to **true**, we mean that the spare element y is assigned to replace the faulty element x.

For example, $W(w) = ab+c$ for vertex w in Example 4.1. This means that vertex w can be repaired either by vertices a and b together, or by vertex c alone. Let $A = \{(w, a), (w, c)\}$ be a subset of edges. Then the evaluation of $W(w)$ with respect to A is $W_A(w) = \mathbf{true} \cdot \mathbf{false} + \mathbf{true} = \mathbf{true}$. This means that faulty element w has been functionally replaced by spare elements a and c.

Notice that negation of variables is not allowed in the expression $W(x)$ since negation of y in $W(x)$ would mean that "vertex y cannot be used to repair vertex x", which corresponds to a situation that has already been captured by the absence of an edge between x and y.

Let y be a vertex in Y. We let $\delta(y) = i$ if vertex y can be used to repair at most i of the vertices in X simultaneously. In particular, $\delta(y) = \infty$ means that vertex y can be used to repair all the vertices in X that are adjacent to it.

Let $P_X = \{\mathcal{X}_1, \mathcal{X}_2, \ldots, \mathcal{X}_l\}$ be a partition on the vertices in X. Let $P_Y = \{\mathcal{Y}_1, \mathcal{Y}_2, \ldots, \mathcal{Y}_r\}$ be a partition on the vertices in Y. For each block $\mathcal{X}_i \in P_X$, there is a non-negative integer, $t(\mathcal{X}_i)$, called the *threshold* of $\mathcal{X}_i$ which indicates the least number of vertices in $\mathcal{X}_i$ that must be repaired by the vertices in Y in order to assure the proper functioning of the chip. For each block $\mathcal{Y}_i \in P_Y$, there is a non-negative integer, $t(\mathcal{Y}_i)$, called the *threshold* of $\mathcal{Y}_i$ which indicates the maximum number of vertices in $\mathcal{Y}_i$ that can be used to repair the vertices in X. The partition P_X captures the situation in which only a certain number of a group of faulty elements need to be repaired, while the partition P_Y captures the situation in which only a certain number of a group of spare elements can be used to repair faulty elements. A partition P_X is said to be *trivial* if $t(\mathcal{X}_i) = |\mathcal{X}_i|$ for each block $\mathcal{X}_i$ in P_X, and is said to be *non-trivial* otherwise. Similarly, a partition P_Y is said to be *trivial* if $t(\mathcal{Y}_i) = |\mathcal{Y}_i|$ for each block $\mathcal{Y}_i$ in P_Y, and is said to be *non-trivial* otherwise. Figure 4.2 shows the graph representing the relationship between the faulty elements and spare elements described in Example 4.1.

For a subset of edges $A \subseteq E$ and a vertex $y \in Y$, *the degree of a vertex y with respect to A*, denoted $d_A(y)$, is defined to be the number of edges in A that are incident with y. A subset of edges $A \subseteq E$ is said to be a *generalized edge cover* if the following three conditions are satisfied:

1. For each block $\mathcal{X}_i \in P_X$, the number of true vertices in $\mathcal{X}_i$ with respect to A is at least the threshold $t(\mathcal{X}_i)$, that is $|\{x \in \mathcal{X}_i : W_A(x) = \textbf{true}\}| \geq t(\mathcal{X}_i)$.

2. For each block $\mathcal{Y}_i \in P_Y$, the number of vertices in the block that are incident with one or more edges of A is at most the threshold $t(\mathcal{Y}_i)$, that is $|\{(x, y) \in A : y \in \mathcal{Y}_i\}| \leq t(\mathcal{Y}_i)$.

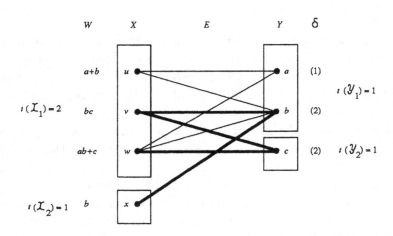

Figure 4.2: The generalized bipartite graph for Example 4.1.

3. For each $y \in Y$, the degree of y with respect to A, $d_A(y)$, is at most $\delta(y)$.

Observe that there is a bijection between generalized edge covers and valid covering assignments. If edge (x, y) is in a generalized edge cover then spare element y is assigned to repair faulty element x in the corresponding covering assignment, and vice versa.

4.3 Illustrative Examples

In this section we present a number of examples to illustrate how various fault covering problems can be represented using the general formulation. The first example is the feasible cover problem for homogeneous arrays studied in Chapter 2. The second example is the feasible cover problem for homogeneous arrays with shared spares. As was noted in Chapter 3, such arrays can increase the efficiency in the use of spares. The third example is the problem of reconfiguring rectangular arrays of processors with interstitial redundancy studied in [76]. Finally, we consider the problem of reconfiguring interstitial arrays in which two distinct types of spares are available.

Recall that in a homogeneous reconfigurable array, spare elements are configured as spare rows and spare columns; a spare row can be used to repair any row of the array and a spare column can be used to repair any column of the array. Figure 4.3a shows a rectangular array consisting of 4 rows and 5 columns of elements together with 2 spare rows and 3 spare columns. Each faulty element x in the array, denoted by a □, is represented by a vertex x in X. Row i and column j of the array are represented by vertices r_i and c_j in Y, respectively. A faulty element x located in row i and column j can be repaired by replacing either row i with a spare row or column j with a spare column. Therefore, there is an edge (x, r_i) and an edge (x, c_j) in E. Also, $W(x) = r_i + c_j$. We let P_X be a trivial partition and let P_Y be a partition with two blocks. The first block, $\mathcal{Y}_1$, consists of the vertices in Y that represent the rows of the array and the threshold of the block is $t(\mathcal{Y}_1) = 2$, since there are two spare rows available in this case. The second block, $\mathcal{Y}_2$, consists of the vertices in Y that represent the columns of the array and the threshold of the block is $t(\mathcal{Y}_2) = 3$, since there are three spare columns available in this case. We let $\delta(y) = \infty$ for all vertices y in Y since a spare row or column can repair all the faults in the line that it replaces. Figure 4.3b shows the generalized bipartite graph representing the array in Figure 4.3a. An example of a generalized edge cover is the set of edges incident with vertices r_2, r_3, c_2, c_3, and c_4.

We next consider a variation of the homogeneous array model in which there are a number of arrays with spares located between adjacent arrays. These spares can be used to replace faulty elements in either one of the two adjacent arrays. Figure 4.4a shows an example where there are nine arrays and six sets of spare rows and six sets of spare columns. The □'s indicate the locations of faulty elements. The feasible cover problem for such arrays can be expressed by the general formulation as shown in Figure 4.4b. The faulty element a, located at column 2 and row 4, can be repaired either by replacing its row with a spare row from D or E or by replacing its column with a spare column from F. Therefore, in the generalized bipartite graph, a is connected to vertices r_{D4}, r_{E4}, and c_{F2}, and $W(a) = r_{D4} + r_{E4} + c_{F2}$. Similar constructions are applied to the other faulty elements. Let $\delta(y) = 1$ for all vertices y in Y, let partition P_X be a trivial partition, and let partition P_Y be a partition in which each block consists of either spare rows or spare columns in any particular position in the array. For example, block $\mathcal{Y}_1$ represents spare rows at location D, block $\mathcal{Y}_2$ represents spare rows at location E, and so

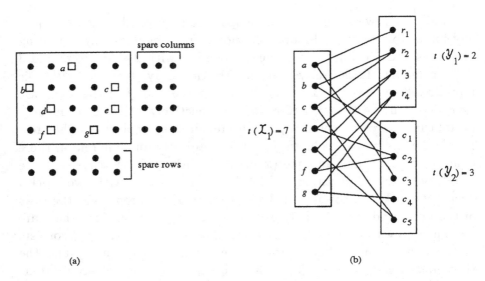

(a)   (b)

Figure 4.3: (a) A faulty homogeneous array. (b) The corresponding graph.

on. The threshold of block $\mathcal{Y}_i$ is the number of spares in the block. For example, $t(\mathcal{Y}_1) = 1$ since there is one spare row at location D.

The third example is a reconfiguration problem for rectangular arrays with interstitial redundancy studied by Singh [76]. In such arrays, spare elements are identical to the regular elements and are placed at the interstitial sites of the array. Each spare is connected to its nearest neighboring regular elements. An instance of such an array is shown in Figure 4.5a where a darkened circle represents a regular element in the array and a spare element is represented by a square. Two overlapping squares represent a spare element that can repair two faulty elements at the same time. In Figure 4.5a a line segment connecting a regular element and a spare element indicates that these two elements are neighbors and the spare element can be used to replace the regular element if it is faulty. Each spare element can replace one faulty element and each faulty element can be replaced by one spare element. Figure 4.5b shows our formulation for the case that elements a, b, and c are faulty. For each spare element y that can repair two faulty elements at the same time, we set $\delta(y) = 2$ and for each spare element z that can repair only one faulty element, we set $\delta(z) = 1$. Finally, we let partitions P_X and P_Y be

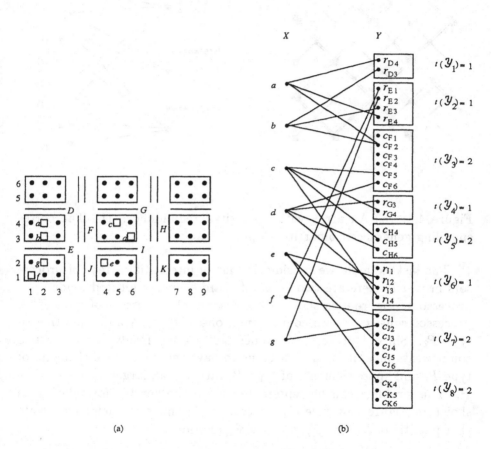

(a) (b)

Figure 4.4: (a) Nine arrays with shared spares. (b) The corresponding generalized bipartite graph.

trivial partitions.

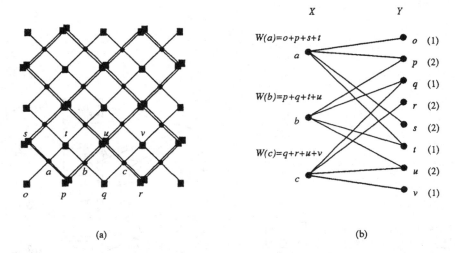

(a)   (b)

Figure 4.5: (a) An array with interstitial redundancy. (b) The corresponding generalized bipartite graph.

The last example we consider is that of interstitial arrays of processors in which there are two kinds of spare elements located among the processors. Figure 4.6a shows a 3×3 array of processors of type P. A processor consists of two components, one of type P_1 and the other of type P_2. Since a processor becomes faulty even if only one of its two components fails, it is advantageous to have separate spare elements of type P_1 and spare elements of type P_2 rather than larger spares of type P. This problem can be represented using the general formulation as shown in Figure 4.6b where processors a, b, and c are defective. Note that partitions P_X and P_Y are trivial partitions.

4.4 Integer Linear Programming Approach

The general formulation presented in Section 4.2 provides a framework for constructing general algorithms for the solution of a large class of fault covering problems. In the next section we shall show that the fault covering problems represented by our formulation are, in general, NP-complete. Therefore, it is very unlikely that there exist polynomial time

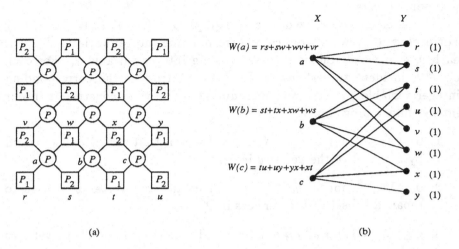

Figure 4.6: (a) Interstitial array with two types of spares. (b) The corresponding generalized bipartite graph.

algorithms for all such fault covering problems. In this section we show how a fault covering problem in the general formulation can be transformed into an integer linear programming problem. Subsection 4.4.1 describes this transformation in detail and and Subsection 4.4.2 provides experimental results for several reconfiguration problems.

4.4.1 The General Transformation

In this section we present a procedure which systematically transforms the problem of finding feasible covers in our general formulation into an integer linear programming problem [25]. As it has been noted, the integer linear programming problem is NP-complete [21]. However, there are several effective methods that can be used to obtain an exact or an approximate solution to the problem. For example, exact solutions can be obtained for problems of moderate size using the cutting-plane algorithm [60] or the branch-and-bound algorithm that computes upper bounds to the solution by solving a set of linear programming problems [61]. For problems of large size, approximate solutions can be obtained. In many special cases, the integer linear programming problem can be reduced to graph theoretic problems that can be solved optimally in polynomial time.

Consider the generalized bipartite graph $G = (X \cup Y, E, W, \delta, P_X, P_Y)$. For each $x_i \in X$ we introduce a corresponding integer variable X_i. For each $y_i \in Y$ we introduce a corresponding integer variable Y_i. For each $e_{ij} \in E$ incident to vertices x_i and y_j, we introduce the corresponding integer variable E_{ij}. We now introduce the following constraints in our integer linear program:

1. $0 \le X_i \le 1$, for all $x_i \in X$. $X_i = 1$ means that vertex x_i is repaired by a subset of the vertices in Y.

2. $0 \le Y_i \le 1$, for all $y_i \in Y$. $Y_i = 1$ means that vertex y_i is used to repair a subset of the vertices in X.

3. $0 \le E_{ij} \le 1$, for all $e_{ij} \in E$. $E_{ij} = 1$ means y_j is one of the vertices used to repair vertex x_i.

For each $x_i \in X$, assume $W(x_i)$ has m product terms, that is,

$$W(x_i) = y_{1,1}y_{1,2} \cdots y_{1,i_1} + y_{2,1}y_{2,2} \cdots y_{2,i_2} + \ldots +$$
$$y_{j,1}y_{j,2} \cdots y_{j,i_j} + \ldots + y_{m,1}y_{m,2} \cdots y_{m,i_m}.$$

We introduce m integer variables T_{ij}, $1 \le j \le m$, corresponding to the m product terms in $W(x_i)$. We now add the following constraints in the integer linear program:

4. $0 \le T_{ij} \le 1$ for all $j \le m$. $T_{ij} = 1$ means that x_i will be repaired by the vertices corresponding to T_{ij}, the jth product term in $W(x_i)$.

5. $T_{i1} + T_{i2} + \ldots + T_{im} \ge X_i$. Thus, if $X_i = 1$ then at least one of T_{ij} must be 1.

6. $E_{i(j,1)} + E_{i(j,2)} + \ldots + E_{i(j,i_j)} - i_j \cdot T_{ij} \ge 0$, for $1 \le j \le m$. Recall that $E_{i(j,l)}$ is the integer variable corresponding to the edge between x_i and $y_{j,l}$. This constraint ensures that all edges between x_i and the vertices in the jth product term of $W(x_i)$ are 1, if the vertices in the jth product term are chosen to repair x_i.

For each block $\mathcal{X}_l$ in the partition P_X, let $x_{l_1}, x_{l_2}, \ldots, x_{l_p}$ be the vertices in the block $\mathcal{X}_l$. We introduce the following constraint to the integer linear program:

7. $X_{l_1} + X_{l_2} + \ldots + X_{l_p} \geq t(\mathcal{X}_l)$. This ensures that the number of vertices repaired in each block is at least the threshold value for that block.

For each $y_i \in Y$, let $x_{i_1}, x_{i_2}, \ldots, x_{i_q}$ be the vertices in X that are adjacent to vertex y_i. We introduce the following constraints to the integer linear program:

8. $E_{i_1 i} + E_{i_2 i} + \ldots + E_{i_q i} \leq \delta(y_i)$. This constraint ensures that vertex y_i is used to repair at most $\delta(y_i)$ vertices in X.

9. $q \cdot Y_i - (E_{i_1 i} + E_{i_2 i} + \ldots + E_{i_q i}) \geq 0$. This is to ensure that if y_i is used then Y_i is equal to 1.

For each block $\mathcal{Y}_j$ in the partition P_Y, let $y_{j_1}, y_{j_2}, \ldots, y_{j_p}$ be the vertices in the block. We introduce the following constraint:

10. $Y_{j_1} + Y_{j_2} + \ldots + Y_{j_p} \leq t(\mathcal{Y}_i)$. This constraint ensures that the number of vertices used in each block of the partition does not exceed the threshold of the corresponding block.

Under such a transformation, it is not difficult to show that a feasible solution to the integer linear programming problem corresponds to a covering assignment for the fault covering problem. Moreover, a minimum feasible solution to the integer linear programming problem corresponds to a covering assignment that minimizes a given objective function. For example, the solution to the integer linear programming problem with an objective function equal to the number of spare elements used corresponds to a covering assignment using the minimum number of spare elements.

As an example, we now transform the fault covering problem in Example 4.1 into an integer linear programming problem. Let variables X_1, X_2, X_3 and X_4 represent vertices u, v, w and x, respectively, in Example 4.1. Similarly, let variables Y_1, Y_2 and Y_3 represent vertices a, b and c, respectively, in Figure 4.1. Following the transformation described above, the constraints are:

1. $0 \leq X_1, X_2, X_3, X_4 \leq 1$

2. $0 \leq Y_1, Y_2, Y_3 \leq 1$

3. $0 \leq E_{11}, E_{12}, E_{22}, E_{23}, E_{31}, E_{32}, E_{33}, E_{42} \leq 1$

4. $0 \leq T_{11}, T_{12}, T_{21}, T_{31}, T_{32}, T_{41} \leq 1$

5. The following constraints ensure that if $X_i = 1$ then at least one of its product terms must also be 1:

   $T_{11} + T_{12} \geq X_1$

   $T_{21} \geq X_2$

   $T_{31} + T_{32} \geq X_3$

   $T_{41} \geq X_4$

6. The following constraints ensure that if the term corresponding to T_{ij} is chosen to repair the vertex x_i then all variables in the jth product term in $W(x_i)$ must be 1:

   $E_{11} - 1 \cdot T_{11} \geq 0$

   $E_{12} - 1 \cdot T_{12} \geq 0$

   $E_{22} + E_{23} - 2 \cdot T_{21} \geq 0$

   $E_{31} + E_{32} - 2 \cdot T_{31} \geq 0$

   $E_{33} - 1 \cdot T_{32} \geq 0$

   $E_{42} - 1 \cdot T_{41} \geq 0$

7. The following constraints correspond to the partition P_X:

   $X_1 + X_2 + X_3 \geq 2$

   $X_4 \geq 1$

8. The following constraints ensure that vertex $y \in Y$ is used to repair at most $\delta(y)$ vertices in X.

   $E_{11} + E_{31} \leq 1$

   $E_{12} + E_{22} + E_{32} + E_{42} \leq 2$

   $E_{23} + E_{33} \leq 2$

9. The following constraints ensure that if any edge adjacent to a vertex y_i is in a generalized edge cover, then Y_i is equal to 1:

   $2 \cdot Y_1 - (E_{11} + E_{31}) \geq 0$

$$4 \cdot Y_2 - (E_{12} + E_{22} + E_{32} + E_{42}) \geq 0$$
$$2 \cdot Y_3 - (E_{23} + E_{33}) \geq 0$$

10. The following constraints correspond to the partition P_Y:

$$Y_1 + Y_2 \leq 1$$
$$Y_3 \leq 1$$

4.4.2 Experimental Results

In this subsection we give experimental results demonstrating the effectiveness of our approach. A program for transforming a generalized bipartite graph into a corresponding integer linear program was implemented in Pascal on a Pyramid computer. The integer linear program was then solved by the LINDO software package on an IBM 3081 computer. Experimental results are given for the minimum feasible cover problems for homogeneous arrays, for homogeneous arrays with shared spares, and for interstitial processor arrays with two types of spares. Generalized bipartite graph formulations were described for these problems in Section 4.3.

Our first experimental results are for the minimum feasible cover problem for homogeneous arrays. Table 4.1 gives results of the integer linear programming approach for test data taken from [46]. Table 4.1 also gives the running times of Kuo and Fuchs' branch-and-bound algorithm [46] on the same data. These experimental results were obtained on different computers and the comparison is made only to provide a reference for the performance of the integer linear programming approach. Table 4.2 gives results for test data taken from [54] with the running times of Lombardi and Huang's algorithm [54] and Reddy and Hemmady's algorithm [32] on the same data. The latter two algorithms were implemented on a VAX 750, and again these comparisons are made only for an approximate comparison. These experimental results indicate that the integer linear programming approach is quite effective, even for large arrays with a large number of faults.

The second example we consider is the minimum feasible cover problem for homogeneous arrays with shared spares. Table 4.3 gives experimental results for 10 randomly generated examples, each consisting of 9 arrays with 12 groups of shared spares as in Figure 4.4. In the largest

Array Size	Spare rows = Spare cols	Faults	Repairable?	Branch-and-Bound time (secs)	Integer Programming time(secs)
128x128	4	5	yes	0.12	0.07
128x128	4	15	no	0.14	0.15
256x256	5	10	yes	0.20	0.09
256x256	5	30	no	0.38	0.20
512x512	5	10	yes	0.28	0.07
512x512	10	19	yes	0.40	0.16
512x512	10	45	no	0.92	0.34
512x512	20	45	yes	1.32	0.39
1024x1024	20	40	yes	1.06	0.34
1024x1024	20	60	no	1.79	0.53
1024x1024	20	200	no	28.26	1.67
1024x1024	20	400	no	178.12	3.47
1024x1024	20	400	yes	1.72	2.33

Table 4.1: Experimental results for homogeneous arrays.

Array Size	Spare rows = Spare cols	Faults	Repairable?	Reddy & Hemmady time (secs)	Lombardi & Huang time (secs)	Integer Programming time(secs)
256x256	4	24	yes	0.07	0.22	0.10
256X256	4	25	yes	0.06	0.20	0.22
256X256	4	28	no	0.02	0.21	0.12
256X256	6	31	yes	0.17	0.25	0.15
512X512	4	27	no	0.08	0.19	0.14
512X512	5	36	no	0.08	0.17	0.14
512X512	7	42	yes	0.17	0.45	0.22
512X512	8	58	yes	0.15	0.41	0.29
1024X1024	4	27	no	0.01	0.17	0.13
1024X1024	5	27	yes	0.14	0.33	0.17
1024X1024	7	61	yes	0.16	0.45	0.30
1024X1024	8	81	no	0.17	0.22	0.38

Table 4.2: Experimental results for homogeneous arrays.

Array Size	Spare rows = Spare cols	Faults	Repairable?	Integer Programming time (secs)
64x64	5	21	yes	0.28
64x64	10	36	yes	0.36
128x128	10	100	no	1.30
128x128	15	113	yes	1.63
256x256	15	139	no	2.07
256x256	20	159	yes	2.41
256x256	20	168	no	3.06
512x512	20	155	yes	2.87
512x512	20	188	no	3.06
512x512	25	204	yes	4.47

Table 4.3: Experimental results for arrays with shared spares.

example, there are nine 512×512 arrays and 25 spare lines in each group. Note again that the computation time is very reasonable, even for large arrays with a large number of faults.

Finally, Table 4.4 gives experimental results for interstitial arrays with two types of spare elements, as described in Section 4.2. These results indicate that our approach is indeed viable for arrays of reasonable dimensions.

4.5 Complexity Analysis of Subcases

In addition to providing a systematic approach to solving a large class of fault covering problems, another advantage of the general formulation is that it enables us to characterize the computational complexity of many of these problems. In Subsection 4.5.1 we identify sixteen subcases of the general formulation together with their complexities. These subcases contain many well-known fault covering problems as special cases. In Subsections 4.5.2 and 4.5.3 we discuss in detail the polynomial time algorithms and the NP-completeness results associated with these

Processor Array Size	Faults	Repairable?	Integer Programming time (secs)
8x8	4	yes	0.29
8x8	8	yes	0.65
16x16	10	yes	1.37
16x16	15	yes	2.68
32x32	20	yes	4.50
32x32	25	yes	6.98
64x64	30	yes	9.70
64x64	35	yes	12.81
128x128	40	yes	16.70
128x128	50	yes	25.79

Table 4.4: Experimental results for interstitial arrays with two types of spares.

subcases.

4.5.1 The Definition of Subcases and Their Complexities

The general formulation provides a framework for classifying fault covering problems. In this subsection, we classify some of the fault covering problems in our general formulation into subcases based on different choices of W, δ, P_X, and P_Y. We consider two choices for each of these parameters, resulting in a total of sixteen subcases. These subcases contain a number of well-known fault covering problems, including all those given as examples in Section 4.3.

In these subcases, the expression $W(x)$ is restricted to be either a *sum* of individual vertices in Y, such as $y_1 + y_2 + y_3$, or a *product* of individual vertices in Y, such as $y_1 y_2 y_3$. For the function δ, we consider the case $\delta(y) = 1$ for all $y \in Y$ and the case $\delta(y) = \infty$ for all $y \in Y$. Finally, we allow P_X to be trivial or non-trivial and P_Y to be trivial or non-trivial. In Table 4.5 we assign names to the sixteen subcases and indicate the complexity of each subcase. For example, the column labeled

	OR $\delta = 1$	*OR* $\delta = \infty$	*AND* $\delta = 1$	*AND* $\delta = \infty$
trivial P_X trivial P_Y	Subcase 1 poly time	Subcase 2 linear time	Subcase 3 linear time	Subcase 4 linear time
non-trivial P_X trivial P_Y	Subcase 5 poly time	Subcase 6 linear time	Subcase 7 NP-complete	Subcase 8 linear time
trivial P_X non-trivial P_Y	Subcase 9 poly time	Subcase 10 NP-complete	Subcase 11 linear time	Subcase 12 linear time
non-trivial P_X non-trivial P_Y	Subcase 13 poly time	Subcase 14 NP-complete	Subcase 15 NP-complete	Subcase 16 NP-complete

Table 4.5: Sixteen subcases of the general formulation.

OR, $\delta = \infty$ in the table refers to the subcases in which the expression $W(x)$ is a sum of the vertices in Y for each vertex $x \in X$, and $\delta(y) = \infty$ for each vertex $y \in Y$. The other three columns are defined in a similar way. For brevity, we refer to the fault covering problem for subcase i as problem i, for $1 \leq i \leq 16$.

In the next two subsections we shall discuss the complexity results in Table 4.5 in detail. In Subsection 4.5.2, we present polynomial time algorithms for solving problems 1, 2, 3, 4, 5, 6, 8, 9, 11, 12, and 13. In Subsection 4.5.3, we show that problems 7, 10, 14, 15, and 16 are NP-complete. Since problem 16 is NP-complete, the problem of finding a generalized edge cover in a generalized bipartite graph in which the expressions $W(x)$ are of the form sum-of-products, P_X and P_Y are non-trivial partitions, and the $\delta(y)$'s are arbitrary positive integers, is also NP-complete.

The subcases in which $W(x)$ is of the form sum-of-products and the partition P_X is trivial can be reduced to subcases in which $W(x)$ is in the form of a single product term and the partition P_X is non-trivial. If $W(x)$ is a sum of α product terms, we can represent the sum by splitting the vertex x into α vertices and associating with each of these vertices one of the product terms in $W(x)$. We then place all these α vertices in a block $\mathcal{X}_i$ in the partition P_X with $t(\mathcal{X}_i) = 1$. An example of this transformation is illustrated in Figure 4.7. Since $W(u) = ab + cd$, u is split into two vertices, called u_1, u_2 with $W(u_1) = ab$ and $W(u_2) = cd$. Similarly, v is split into vertices v_1 and v_2 with $W(v_1) = b$ and $W(v_2) = c$, and w is split into vertices w_1 and w_2 with $W(w_1) = c$ and $W(w_2) = d$.

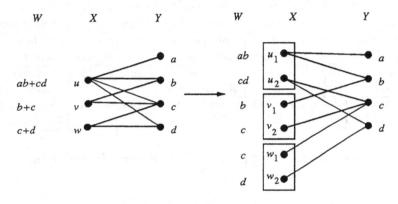

Figure 4.7: Splitting vertices u, v, and w.

4.5.2   Polynomial Time Algorithms

We shall show that problems 1, 2, 3, 4, 5, 6, 8, 9, 11, 12, and 13 can be solved in polynomial time. These problems can be partitioned into two classes. The first class, consisting of problems 2, 3, 4, 6, 8, 11, and 12, can be solved in linear time in terms of the number of vertices and edges in the generalized bipartite graph. The second class, consisting of problems 1, 5, 9, and 13, can be solved in polynomial time by reductions to network flow problems. We discuss these two classes separately.

Problems 2, 3, 4, 6, 8, 11, and 12 can be solved in time linear in the number of vertices and edges in the generalized bipartite graph by a simple traversal of the vertices in the graph. In problems 2, 4, 6, and 8, all the vertices in Y can be used to repair the vertices in X. Thus, we can examine each vertex $x \in X$ to see if this results in a covering assignment. If this is a covering assignment then the chip is repairable and is unrepairable otherwise. In problems 3, 11, and 12 we must repair all faulty elements. Thus, we can simply scan each vertex $x \in X$ and see if it can be repaired by one or more of the vertices in Y. If that is the case, and the solution is a covering assignment, then the chip is repairable. Otherwise, the chip is unrepairable.

Problems 1, 5, 9, and 13 can be solved by network flow algorithms. We show how problem 13 can be transformed into a generalized maximum flow problem in which there is a lower and an upper bound on the flow value associated with each of the edges. Such a generalized maximum flow problem can then be reduced in linear time to the maximum flow problem as shown in [53]. The maximum flow problem can be solved by any of a number of well-known algorithms. For example, Goldberg and Tarjan's [22] algorithm runs in time $O(nm\log(n^2/m))$ where n is the number of vertices and m is the number of edges in the network. The same transformation can be applied to problems 1, 5, and 9 with only minor modifications.

Recall that in problem 13, each faulty element can be repaired by one of several spare elements and each spare element can repair no more than one faulty element. Let $G = (X \cup Y, E, W, \delta, P_X, P_Y)$ be a generalized bipartite graph representing an instance of problem 13. First, a directed graph $G' = (V', E')$ is constructed from G. The construction, illustrated in Figure 4.8, is as follows:

1. All vertices in the generalized bipartite graph G belong to V'.

2. All edges in the generalized bipartite graph G belong to E' with their direction being from vertices in X to vertices in Y. The lower and upper bounds on the values of the flows in these edges are 0 and 1, respectively.

3. For each block $\mathcal{X}_i$ of P_X, a vertex $\bar{x}_i$ is added to V' to represent the block $\mathcal{X}_i$. There are directed edges from $\bar{x}_i$ to all vertices in $\mathcal{X}_i$. The lower and upper bounds on the values of the flows in these edges are 0 and 1, respectively.

4. For each block $\mathcal{Y}_i$ of P_Y, a vertex $\bar{y}_i$ is added to V' to represent the block $\mathcal{Y}_i$. There are directed edges from $\bar{y}_i$ to all the vertices in $\mathcal{Y}_i$. The lower and upper bounds on the values of the flows in these edges are 0 and 1, respectively.

5. A distinct vertex s, called the *source*, is added to V'. There are directed edges from s to all vertices $\bar{x}_i$. The lower bound on the value of the flow in an edge between the source s and the vertex $\bar{x}_i$ is $t(\mathcal{X}_i)$ and the upper bound is $|\mathcal{X}_i|$.

6. A distinct vertex z, called the *sink*, is added to V'. There are directed edges from all vertices $\bar{y}_j$ to z. The lower bound on the value of the flow in an edge between $\bar{y}_i$ and the sink z is 0 and the upper bound is $t(\mathcal{Y}_j)$.

Since G' has integer upper and lower bounds on the flow values, there is an integer feasible flow in G' [85]. We now show that an integer feasible flow in G' corresponds to a generalized edge cover, and thus a covering assignment, in G. Suppose there is a feasible flow in G'. We note that the edges from X to Y in which the value of the flow is 1 constitute a generalized edge cover for G. This is based on the following observations: The flow value from the source to each vertex $\bar{x}_i$ is at least the threshold of block $\mathcal{X}_i$. Thus, at least $t(\mathcal{X}_i)$ of the vertices in block $\mathcal{X}_i$ will have an incoming flow of value 1. Also, exactly one of the outgoing edges of each of these vertices will have its flow value equal to 1. Furthermore, at most $t(\mathcal{Y}_i)$ vertices in each block $\mathcal{Y}_i$ will have an incoming flow of value 1, since the maximum flow from each vertex $\bar{y}_i$ to the sink is $t(\mathcal{Y}_i)$.

Conversely, suppose there is a generalized edge cover in G. For each edge (u, v) in the generalized edge cover, let $(\hat{u}, \hat{v})$ be the corresponding directed edge in G'. Set the value of the incoming flow to $\hat{u}$, the value of outgoing flow from $\hat{u}$, and the value of outgoing flow from $\hat{v}$ to 1. Let the value of the incoming flow to $\bar{x}_i$ be the sum of the outgoing flow values from $\bar{x}_i$ and let the value of outgoing flow from $\bar{y}_i$ be the sum of the incoming flow value to $\bar{y}_i$. It can be verified that this construction will result in an integer feasible flow in G'.

Table 4.6 shows some experimental results for problem 13 using Goldberg and Tarjan's algorithm implemented in Pascal on a Pyramid computer. The faults were generated randomly.

4.5.3 NP-Completeness Results

The feasible cover problem for homogeneous arrays was shown to be NP-complete by Kuo and Fuchs [46], as was discussed in Chapter 2.

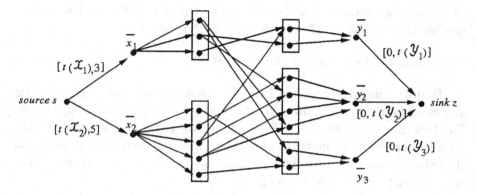

Figure 4.8: Network construction for problem 13.

Defective Elements	Redundant Elements	Blocks in P_X	Blocks in P_Y	Repairable?	Network Flow time (secs)
26	22	7	7	yes	.383
32	32	5	5	yes	.583
50	48	7	6	no	6.116
76	64	7	7	no	11.016
82	79	11	8	yes	1.666
95	82	8	8	yes	2.100
124	107	11	14	no	24.250
194	157	18	16	yes	6.216

Table 4.6: Experimental results for problem 13.

This problem is a special case of problem 10, and thus problem 10 is also NP-complete. Consequently, problem 14 is also NP-complete. In the following theorem, we show that problem 7 is NP-complete, which implies that problems 15 and 16 are both NP-complete.

Theorem 17 *Problem 7 is NP-complete.*

Proof: First, the problem is in NP since we can guess a set of edges and check in polynomial time whether this set is a generalized edge cover. To show that problem 7 is NP-complete, we reduce an instance of the 3-SAT problem to an instance of problem 7 in polynomial time and show that the instance of the 3-SAT problem has a solution if and only if the instance of problem 7 has a solution.

Recall that an instance of the 3-SAT problem can be stated as follows: given a set of boolean variables $U = \{u_1, u_2, \ldots, u_k\}$, of variables, a set of clauses $\mathcal{D} = \{D_1, D_2, \ldots, D_n\}$ each consisting of the disjunction of exactly three variables from U, and the conjunction of these clauses, S, does there exist an assignment that satisfies S? Given an instance of 3-SAT, we now construct an instance of problem 7.

The generalized bipartite graph $G = (X \cup Y, E, W, \delta, P_X, P_Y)$ is constructed as follows. Vertex set X contains vertices x_{ij} and $\bar{x}_{ij}$ for each i and j if variable u_i or $\neg u_i$ appears in clause D_j. Vertex set X also contains vertices w_{j1}, w_{j2}, w_{j3} for each clause D_j. Vertex set Y contains vertices $y_{ij}, \bar{y}_{ij}$, and z_{ij} for each i and j if variable u_i or $\neg u_i$ appears in D_j. In addition, there are four kinds of edges in E. They are:

1. $(x_{ij}, y_{ij}) \in E$ and $(\bar{x}_{ij}, \bar{y}_{ij}) \in E$.

2. $(x_{ij}, z_{ij'}) \in E$, where either:

   (a) $j' < j$ and there is no $\hat{j}$, $j' < \hat{j} < j$, such that there is a vertex $z_{i\hat{j}}$ in Y, or

   (b) there is no $\hat{j} < j$ such that there is a vertex $z_{i\hat{j}}$ in Y, and $j' > \hat{j}$ for all $\hat{j}$ such that there is a vertex $z_{i\hat{j}}$.

3. $(\bar{x}_{ij}, z_{ij}) \in E$.

4. $(w_{jk}, y_{ij}) \in E$ or $(w_{jk}, \bar{y}_{ij}) \in E$ depending on whether u_i or $\bar{u}_i$ is the kth variable in D_j, $k = 1, 2,$ or 3.

For each $x \in X$, the expression $W(x)$ is a product of the vertices in Y that are adjacent to x. For all i and j, $\delta(y_{ij}) = 1$, $\delta(\bar{y}_{ij}) = 1$ and $\delta(z_{ij}) = 1$. The partition on Y is a trivial partition. The partition on X consists of the following blocks. For each i and j, x_{ij} and $\bar{x}_{ij}$ form a block with threshold equal to 1. For each j, w_{j1}, w_{j2}, w_{j3} also form a block with threshold equal to 1.

Note that for every variable u_i, either all edges of the form $(x_{ij}, z_{ij'})$ are included in and all edges of the form $(\bar{x}_{ij}, z_{ij''})$ are excluded from a generalized edge cover or all edges of the form $(x_{ij}, z_{ij'})$ are excluded from and all edges of the form $(\bar{x}_{ij}, z_{ij''})$ are included in a generalized edge cover. This is the case because each z_{ij} can repair only one vertex of X. This means that only the y_{ij}'s or the $\bar{y}_{ij}$'s are available to repair the w_{kj}'s.

If there is a solution to the given instance of 3-SAT, then we can derive a generalized edge cover as follows. For every variable u_i of clause D_j that is set to **true**, we let edge (w_{jk}, y_{ij}) be in the generalized edge cover, where $k = 1, 2$, or 3 and u_i is the kth variable in D_j. The two edges incident with each $\bar{x}_{ij}$, for all clauses D_j that contain either u_i or $\neg u_i$, are also in the generalized edge cover. For every variable $\neg u_i$ of clause D_j that is set to **true**, we let edge $(w_{jk}, \bar{y}_{ij})$ be in the generalized edge cover, where $k = 1, 2$, or 3 and u_i is the kth variable in D_j. The two edges incident with each x_{ij}, for all clauses D_j that contain either u_i or $\neg u_i$, are also in the generalized edge cover. All other edges are not in the generalized edge cover. Note that if variable u_i is set to **true** in D_j then it is also set to **true** in all other clauses. This means that either each y_{ij} is available to repair x_{ij} or each $\bar{y}_{ij}$ is available to repair $\bar{x}_{ij}$. This is indeed a generalized edge cover because there is at least one vertex in every block of the partition of X that is repaired and each vertex in Y repairs at most one vertex of X.

Conversely, if there is a solution to the instance of problem 7, we can derive a truth assignment to the variables as follows. For each y_{ij} that is used to repair w_{jk}, we set u_i to **true** and for each $\bar{y}_{ij}$ that is used to repair w_{jk}, we set $\neg u_i$ to **true**. The rest of the variables are set to **false**. As noted earlier, a generalized edge cover will have either each x_{ij} repaired by y_{ij} or each $\bar{x}_{ij}$ repaired by $\bar{y}_{ij}$, for all j. This implies that the **true** assignment of the variables will be consistent in all clauses. Since at least one of w_{jk}, $k = 1, 2$, or 3, is repaired, each clause D_j will have

at least one **true** variable, so this is a solution to the instance of 3-SAT.
□

As an example of this construction, consider the the following instance of 3SAT: $S = (u_1 + \neg u_2 + u_3)(\neg u_1 + u_2 + \neg u_3)(\neg u_1 + \neg u_2 + u_3)$. The corresponding generalized bipartite graph G is shown in Figure 4.9. Edges of type 1 are shown in thin lines, edges of type 2 are shown in thin dotted lines, edges of type 3 are shown in thick dotted lines, and edges of type 4 are shown in thick lines. Finally, we have the following corollary:

Corollary 2 *Problems 15 and 16 are NP-complete.*

4.6 Summary

In this chapter we presented a general and uniform formulation of fault covering problems for different classes of reconfigurable chips. This general formulation provides a framework for constructing algorithms for the solution of a large class of fault covering problems. In addition, this formulation enables us to characterize the computational complexity of a large class of fault covering problems.

After describing the general formulation, we gave a systematic procedure which transforms a fault covering problem in the general formulation into an integer linear programming problem. The integer linear programming problem is a well-studied combinatorial optimization problem for which a number of algorithms and heuristics are known. We demonstrated the effectiveness of our approach by solving three important fault covering problems. The experimental results for these problems were indeed quite good.

Next, we used this general formulation to study the complexity of several classes of fault covering problems. We gave a complete characterization of the complexity of the fault covering problems in each of the sixteen subcases. These subcases contain many important fault covering problems. Simple linear time algorithms were presented for several subcases while it was observed that several others can be solved in polynomial time by reduction to the network flow problem. We showed that the remaining subcases are NP-complete.

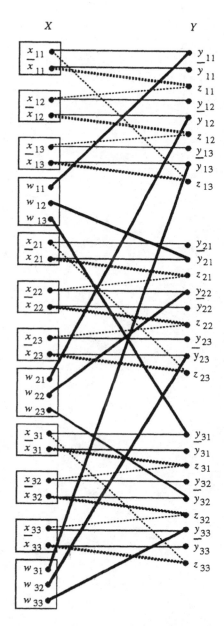

Figure 4.9: An example of the construction in Theorem 16.

Bibliography

[1] *Topical Meeting on Three Dimensional Integration*, Miyagi-Zao, Japan, May 30-June 1, 1988.

[2] V. K. Agarwal. Multiple fault detection in programmable logic arrays. *IEEE Transactions on Computers*, C-29(6):518–522, June 1980.

[3] P. Banerjee. The cubical ring connected cycles: A fault-tolerant parallel computation network. *IEEE Transactions on Computers*, pages 632–636, May 1988.

[4] K. E. Batcher. Design of a massively parallel processor. *IEEE Transactions on Computers*, C-29(9):836–840, Sept. 1980.

[5] C. A. Benevit et al. 256K random access memory. *IEEE Journal on Solid State Circuits*, SC-17(5):857–861, Oct. 1982.

[6] A. A. Bertossi and M. A. Bonuccelli. A gracefully degradable VLSI system for linear programming. *IEEE Transactions on Computers*, 38(6):853–861, June 1989.

[7] M. Chean and J. A. B. Fortes. A taxonomy of reconfiguration techniques for fault-tolerant processor arrays. *IEEE Computer*, 23(1):55–69, Jan. 1990.

[8] M. Chean and J.A.B. Fortes. FUSS: A reconfiguration scheme for fault-tolerance processor arrays. *Int. Workshop on Hardware Fault Tolerance in Multiprocessors*, pages 30–32, June 1989.

[9] A. Chen. Redundancy in LSI memory arrays. *IEEE J. Solid-State Circuits*, pages 291–293, Oct. 1969.

[10] M. Davis and H. Putnam. A computing procedure for quantification theory. *J. ACM*, 7:201–215, 1960.

[11] R. J. Day. A fault-driven comprehensive redundancy algorithm. *IEEE Design and Test*, Vol. 2(3):35–44, June 1985.

[12] M. Demjanenko and S. Upadhyaya. Dynamic techniques for yield enhancement of field programmable logic arrays. In *1988 IEEE Int. Test Conf.*, pages 485–491, 1988.

[13] A. Despain and D. Patterson. X-tree: A tree structured multiprocessor computer architecture. *Proc. 5th Annual Symposium on Computer Architecture*, pages 144–151, April 1978.

[14] S. Dutt and J. P. Hayes. Design and reconfiguration strategies for near-optimal k-fault-tolerant tree architectures. *Proceedings of the 18th International Symposium on Fault-Tolerant Computing*, pages 328–333, 1988.

[15] A. V. Ebel et al. A NMOS 64K static RAM. In *Proc. IEEE Int. Solid State Circuits Conf.*, pages 254–255, Feb. 1982.

[16] J. Edmonds. Paths, trees, and flowers. *Canad. J. Math.*, 17:449–467, 1965.

[17] R. Evans. Testing repairable RAMs and mostly good memories. In *Proc. IEEE Int. Test Conference*, pages 49–55, 1981.

[18] S. Even, A. Itai, and A. Shamir. On the complexity of timetable and multicommodity flow problems. *SIAM J. Computing*, 5:691–703, 1976.

[19] B. F. Fitzgerald and E. P. Thoma. A 288K dynamic RAM. In *Proc. IEEE Int. Solid State Circuits Conf.*, pages 68–69, Feb. 1982.

[20] B. F. Fitzgerald and E. P. Thoma. Circuit implementation of fusible redundant addresses on RAMs for productivity enhancement. *IBM J. Res. Develop.*, 24(3):291–298, May 1980.

[21] M. R. Garey and D. S. Johnson. *Computers and Intractability: A Guide to the Theory of NP-Completeness*. W. H. Freeman and Company, New York, 1979.

[22] A. V. Goldberg and R. E. Tarjan. A new approach to the maximum flow problem. In *Proc. Symposium on Theory of Computation*, pages 136–146, 1986.

[23] R. Gupta, A. Zorat, and I. V. Ramakrishnan. A fault-tolerant multi-pipeline architecture. In *Proceedings of the 16th International Symposium on Fault-Tolerant Computing*, pages 350–355, 1986.

[24] N. Hasan, J. Cong, and C. L. Liu. A new formulation of yield enhancement problems for reconfigurable chips. In *Proc. Conf. Computer-Aided Design*, pages 520–523, Nov. 1988.

[25] N. Hasan, J. Cong, and C. L. Liu. An integer linear programming approach to general fault covering problems. In *Proc. IEEE Int. Workshop on Defect and Fault Tolerance in VLSI Systems*, Oct. 1989.

[26] N. Hasan and C. L. Liu. Minimum fault coverage in reconfigurable arrays. In *Proceedings of the 18th International Symposium on Fault-Tolerant Computing*, pages 27–30, 1988.

[27] N. Hasan and C. L. Liu. Fault covers in reconfigurable PLAs. In *Proceedings of the 20th International Symposium on Fault-Tolerant Computing*, pages 166–173, 1990.

[28] A. S. Hassan and V. K. Agarwal. A fault-tolerant modular approach architecture for binary trees. *IEEE Transactions on Computers*, C-35(4):356–361, April 1986.

[29] Y. Hayasaka, K. Shimotori, and K. Okada. Testing system for redundant memory. In *Proceedings of the IEEE International Test Conference*, pages 240–244, 1982.

[30] J. P. Hayes. A graph model for fault-tolerant computing systems. *IEEE Transactions on Computers*, C-25(9):875–884, Sept. 1976.

[31] I. Heller and C. Tompkins. An extension of a theorem of Dantzig's. In H. Kuhn and A. Tucker, editors, *Linear Inequalities and Related Systems*, pages 247–252. Princeton University Press, 1956.

[32] V. Hemmady and S. Reddy. On the repair of redundant RAMs. In *Proc. 26th Design Automation Conf.*, pages 710–713, June 1989.

[33] A. Hoffman and J. Kruskal. Integral boundary points of convex polyhedra. In H. Kuhn and A. Tucker, editors, *Linear Inequalities and Related Systems*, pages 223–246. Princeton University Press, 1956.

[34] J. Hopcroft and R. M. Karp. An $n^{5/2}$ algorithm for maximum matching in bipartite graphs. *SIAM J. Computing*, 2(4):225–231, 1973.

[35] M. Howells and V. Agarwal. Yield and reliability enhancement of large area binary tree architectures. *Proceedings of the 17th International Symposium on Fault-Tolerant Computing*, pages 290–295, 1987.

[36] M. Howells and V. Agarwal. A reconfiguration scheme for yield enhancement of large array binary tree architectures. *IEEE Transactions on Computers*, C-37(4):463–468, April 1988.

[37] H. L. Kalter et al. An experimental 80ns 1Mb CMOS DRAM with fast page operation. In *Proc. IEEE Int. Solid State Circuits Conf.*, pages 248–249, Feb. 1985.

[38] N. Karmarkar. A new polynomial time algorithm for linear programming. *Combinatorica*, 4:373–395, 1984.

[39] R. A. Kertis et al. A 59ns 256K DRAM using LD^3 technology and double level metal. In *Proc. IEEE Int. Solid State Circuits Conf.*, pages 96–97, Feb. 1984.

[40] J. H. Kim and S. M. Reddy. On the design of fault-tolerant two-dimensional systolic arrays for yield enhancement. *IEEE Transactions on Computers*, 38(4):501–514, April 1989.

[41] K. Kokkonen et al. Redundancy techniques for fast static RAMs. In *Proc. Int. Solid State Circuits Conf.*, pages 80–81, Feb. 1981.

[42] I. Koren. A reconfigurable and fault-tolerant multiprocessor array. *Proc. 8th Annual Symposium on Computer Architecture*, pages 425–442, 1981.

[43] M. Kumanoya et al. A 90ns 1Mb DRAM with multi bit test mode. In *Proc. IEEE Int. Solid State Circuits Conf.*, pages 240–241, Feb. 1985.

[44] H. T. Kung and M. S. Lam. Wafer-scale integration and two-level pipelined implementations of systolic arrays. *J. Parallel Distributed Computing*, pages 32–63, 1984.

[45] S. Kung, S. Jean, and C. Chang. Fault-tolerant array processors using single-track switches. *IEEE Transactions on Computers*, 38(4):501–514, April 1989.

[46] S. Y. Kuo and W. K. Fuchs. Efficient spare allocation for reconfigurable arrays. *IEEE Design and Test*, 4(1):24–31, Feb. 1987.

[47] C. L. Kwan and S. Toida. An optimal 2-fault tolerant realization of symmetric hierarchical tree systems. *Networks*, 12:231–239, 1982.

[48] C. W. H. Lam, H. F. Li, and R. Jayakumar. A study of two approaches for reconfiguring fault-tolerant systolic arrays. *IEEE Transactions on Computers*, 38(6):833–844, June 1989.

[49] H. F. Li, R. Jayakumar, and C. Lam. Restructuring fault-tolerant systolic arrays. *IEEE Transactions on Computers*, page 307, Feb. 1989.

[50] R. Libeskind-Hadas. *Reconfiguration Problems for VLSI Systems*. PhD thesis, University of Illinois at Urbana-Champaign, In preparation.

[51] W. Lin and C. L. Wu. A fault-tolerant mapping scheme for a configurable multiprocessor system. *IEEE Transactions on Computers*, page 227, Feb. 1989.

[52] J. R. Lineback. Vertical shorts add row to memory. *Electronics*, pages 50–52, Sept. 8 1982.

[53] C. L. Liu. *Introduction to Combinatorial Mathematics*. McGraw-Hill, New York, 1968.

[54] F. Lombardi and W. K. Huang. Approaches for the repair of VLSI/WSI RRAMs by row/column deletion. In *Proceedings of the 18th International Symposium on Fault-Tolerant Computing*, pages 342–347, June 1988.

[55] M. B. Lowrie and W. K. Fuchs. Reconfigurable tree architectures using subtree oriented fault tolerance. *IEEE Transactions on Computers*, C-36(10):1172–1182, Oct. 1987.

[56] T. Mano, M. Wada, N. Ieda, and M. Tanimoto. A redundancy circuit for a fault-tolerant 256K MOS RAM. *IEEE Journal on Solid State Circuits*, SC-17(4):726–731, Aug. 1982.

[57] J. V. McCanny and J. G. McWhirter. Yield enhancement of bit level systolic array chips using fault tolerant techniques. *Electronics Letters*, 19(14):525–527, July 7, 1983.

[58] T. Nakano et al. A sub 100ns 256Kb DRAM. In *Proc. IEEE Int. Solid State Circuits Conf.*, pages 224–225, Feb. 1983.

[59] Y. Nishimura, M. Hamada, H. Hidaka, H. Ozaki, and K. Fujishima. A redundancy test-time reduction technique in 1-Mbit DRAM with a multibit test mode. *IEEE Journal on Solid State Circuits*, 24(1):43–49, Feb. 1989.

[60] C. Papadimitriou and K. Steiglitz. *Combinatorial Optimization: Algorithms and Complexity*. Prentice-Hall, Englewood Cliffs, NJ, 1982.

[61] R. G. Parker and R. L Rardin. *Discrete Optimization*. Academic Press, New York, 1988.

[62] F. P. Preparata and J. Vuillemin. The cube-connected cycles, a versatile network for parallel computation. *Communications of the ACM*, pages 30–39, May 1981.

[63] C. S. Raghavendra, A. Avizienis, and M. Ercegovac. Fault tolerance in binary tree architectures. *IEEE Transactions on Computers*, C-33(6):568–572, June 1984.

[64] D. Rinerson et al. 512K EPROMS. In *Proc. IEEE Int. Solid State Circuits Conf.*, pages 136–137, Feb. 1984.

[65] A. Rosenberg. A hypergraph model for fault-tolerant VLSI processor arrays. *IEEE Transactions on Computers*, pages 578–584, June 1985.

[66] A. Rosenberg. The Diogenes approach to testable fault-tolerant networks of processors. *IEEE Transactions on Computers*, C-32(10):902–910, Oct. 1983.

[67] V. P Roychowdhury, J. Bruck, and T. Kailath. Efficient algorithms for reconfiguration in VLSI/WSI arrays. *IEEE Trans. on Computers*, 39(4):480–489, Apr. 1990.

[68] M. Sami and R. Stefanelli. Fault-tolerance and functional reconfiguration in VLSI arrays. *Proc. Int. Conf. Circuits Syst.*, pages 643–648, 1986.

[69] M. Sami and R. Stefanelli. Reconfigurable architectures for VLSI processing arrays. *Proc. IEEE*, pages 712–722, May 1986.

[70] M. G. Sami and R. Stefanelli. Reconfigurable architectures for VLSI implementation. *Proc. NCC 83, AFIPS*, May 1983.

[71] S. E. Schuster. Multiple word/bit line redundancy for semiconductor memories. *IEEE J. Solid-State Circuits*, SC-13(5):698–703, Oct. 1978.

[72] Y. N. Shen and F. Lombardi. Location and identification for single and multiple faults in testable redundant PLAs for yield enhancement. In *IEEE Int. Test Conf.*, pages 528–535, April 1988.

[73] W. Shi, M. F. Chang, and K. Fuchs. On repairing reconfigurable memory arrays. In preparation.

[74] L. A. Shombert and D. P. Siewiorek. Using redundancy for concurrent testing and repairing of systolic arrays. In *Proceedings of the 17th International Symposium on Fault-Tolerant Computing*, pages 244–247, 1987.

[75] A. Singh. A reconfigurable modular fault tolerant binary tree architecture. *Proceedings of the 17th International Symposium on Fault-Tolerant Computing*, pages 298–304, 1987.

[76] A. D. Singh. Interstitial redundancy: An area efficient fault tolerance scheme for large area VLSI processor arrays. *IEEE Transactions on Computers*, C-37(11):1398–1410, Nov. 1988.

[77] R. J. Smith et al. 32K and 16K static MOS RAMs using laser redundancy techniques. In *Proc. IEEE Int. Solid State Circuits Conf.*, pages 252–253, Feb. 1982.

[78] R. T. Smith. Using a laser beam to substitute good cells for bad. *Electronics*, pages 131–134, Jul. 28, 1981.

[79] R. T. Smith et al. Laser programmable redundancy and yield improvement in a 64K DRAM. *IEEE Journal on Solid State Circuits*, SC-16(5):506–513, Oct. 1981.

[80] L. Snyder. Introduction to the configurable, highly parallel computer. *IEEE Computer*, 15:47–56, Jan. 1982.

[81] C. H. Stapper. Yield models for fault clusters within integrated circuits. *IBM J. Res. Develop.*, 28:636–640, Sept. 1984.

[82] C. H. Stapper, A. N. McLaren, and M. Dreckmann. Yield model for productivity optimization of VLSI memory chips with redundancy and partially good product. *IBM J. Res. Develop.*, 24(3):398–409, May 1980.

[83] R. Sud and K. C. Hardee. Designing static RAMs for yield as well as speed. *Electronics*, pages 121–126, Jul. 28, 1981.

[84] E. Tammaru and J. Angell. Redundancy for LSI yield enhancement. *IEEE J. Solid-State Circuits*, SC-2:172–182, Dec. 1967.

[85] R. E. Tarjan. *Data Structures and Network Algorithms*. Society for Industrial and Applied Mathematics, Philadelphia, 1983.

[86] M. Tarr, D. Boudreau, and R. Murphy. Defect analysis system speeds test and repair of redundant memories. *Electronics*, pages 175–179, Jan. 12, 1984.

[87] R. Taylor and M. Johnson. A 1Mb CMOS DRAM with a divided bitline matrix architecture. In *Proc. IEEE Int. Solid State Circuits Conf.*, pages 242–243, Feb. 1985.

[88] J. Tyszer. A multiple fault-tolerant processor network architecture for pipeline computing. *IEEE Transactions on Computers*, pages 1414–1418, Nov. 1988.

[89] P. Vaidya. An algorithm for linear programming which requires $O(((m + n)^2 + (m + n)^{1.5}n)L)$ arithmetic operations. In *Proc. 19th Annual ACM Symp. on Theory of Computing*, pages 29–38, 1987.

[90] P. J. Varman and I. V. Ramakrishnan. Optimal matrix multiplication on fault tolerant VLSI arrays. *IEEE Transactions on Computers*, page 278, Feb. 1989.

[91] C. L. Wey. On yield considerations for the design of redundant programmable logic arrays. *IEEE Trans. on Computer-Aided Design*, 27(4):528–535, April 1988.

[92] C. L. Wey and F. Lombardi. On the repair of redundant RAMs. *IEEE Trans. on Computer-Aided Design*, 6(2):222–231, March 1987.

[93] C. L. Wey, M. K. Vai, and F. Lombardi. On the design of a redundant programmable logic array. *IEEE J. Solid-State Circuits*, 22(1):114–117, Feb. 1987.

[94] M. W. Yung, M. J. Little, R. D. Etchells, and J. G. Nash. Redundancy for yield enhancement in the 3-D computer. In *Proc. 1989 IEEE Int. Conf. on Wafer Scale Integration*, pages 73–82, 1989.

Index

Admissible assignment 21, 23
    maximum, 23
Admissible set 23
Alternating path 27
Augmenting path 27

BF-k critical sets 53
Binary trees 5, 6, 8
Bipartite graph representation 25
Bit line 10
Blocks 69
Bridging faults 50

CHiP 4
Cluster-proof scheme 6
Complex-stealing 4
Constrained vertex cover 52
Constraint matrix 65
Cost function 33
Cousin-connected tree (CCT) 5
Cover algorithm 45
Covering approach 2
Covering assignment 2
Cover 20
Critical set algorithm 24
Critical sets 24, 25
Crosspoint faults 50
Cube-connected cycles (CCC) 6
Cut 5

Data flow paths 5
Deficiency 27
Derivable 21
Diogenes approach 6, 7
Direct reconfiguration 4
Disjoint covers 57
Dynamic reconfiguration 2

Embedding approach 2
Embed 3
Evaluation 94
Excess-k Cover algorithm 45
Excess-k critical sets 39
Excess-k critical sets 41
Expansion 21

Faulty array 7
Fault-stealing
    complex-stealing, 4
    fixed-stealing, 4
    variable-stealing, 4
Feasible coloring 59
Feasible flow 112
Feasible cover problem
    general array model, 77, 78
    heterogeneous array model,
        57, 58
    homogeneous array model, 20
Feasible minimum cover 20
Feasible minimum cover problem

general array model, 77
heterogeneous array model,
  57, 62
homogeneous array model, 20
Feasible minimum vertex cover
  27
Feasible set of covers
  heterogeneous array model,
  57
Feasible vertex cover 27
Fixed-stealing 4
Full use of suitable spares (FUSS)
  5
Fusible links 12
  electrical, 14, 15
  laser 14

General array model 56, 76, 78
Generalized bipartite graph 94
Generalized edge cover 95

Hard switches 2
Heterogeneous array model 56
Homogeneous array model 55, 96

Incremental exhaustive search al-
  gorithm 21
Index-mapping 3
Integer linear programming 67,
  100
Interstitial redundancy 9, 96, 98,
  100
König-Egerváry Theorem 28

LINDO 105
Linear arrays 3, 4
Linear arrays 4
Linear program 65

Line 20

Matched edges 26
Matched vertex 26
Matching 26
Maximum admissible assignment
  23
Maximum deficiency set 27
Maximum matching 26
Min-Cover algorithm 33
Min-Cover algorithm 38
Minimum cover 20
Minimum feasible cover problem
  general array model, 77, 78
  heterogeneous array model,
  57, 64, 71
  homogeneous array model, 41
Minimum feasible cover decision
  problem
  for general arrays, 86
  for heterogeneous arrays, 68
Minimum vertex cover 27
MPP computer 8
Multiple pipelinesd 4

Network flow 110
Non-trivial partition 95

Optimal solution 65

Partial solution 20
Path 27
  alternating, 27
  augmenting, 27
  vertex disjoint, 27
Perfect matching 26
Pipelines 4

Range 27

Reconfigurable systems 1
Reconfiguration problem 1
Redundant programmable logic
        arrays (RPLAs) 50
Reliability 2
Residual array 25
Roving spares 4

Shared spare feasible cover prob-
        lem 49, 97
SOFT 6
Soft switches 2
Solution 19
Spare column array 56, 76
Spare row array 56, 76
Static reconfiguration 2
Stuck-at faults 50
Surplus vector 5
Switch
     hard, 2
     soft, 2
Systolic arrays 5, 8

Three-dimensional VLSI 74
3-satisfiability (3SAT) 82
3-D computer 9
Threshold 95
Totally unimodular (TUM) ma-
        trix 65
Trivial partition 95
Two-dimensional arrays 4, 5
2-satisfiability (2SAT) 58, 62
Type
     column, 76
     row, 76

Unimodular (UM) matrix 65
Unmatched edges 26

Unmatched vertex 26

Variable-stealing 4
Vertex cover 27
        bipartite graph, 62
        minimum, 27
Vertex cover problem 69
Vertex disjoint paths 27

Word line 9

X-tree topology 5
Yield enhancement 2